Technology Assessment

edited by
Raphael G. Kasper

Published in cooperation with the Program of Policy Studies in Science and Technology, George Washington University

Technology Assessment

Understanding the Social Consequences of Technological Applications

PRAEGER SPECIAL STUDIES IN U.S. ECONOMIC AND SOCIAL DEVELOPMENT

Praeger Publishers New York Washington London

PRAEGER PUBLISHERS
111 Fourth Avenue, New York, N.Y. 10003, U.S.A.
5, Cromwell Place, London S.W.7, England

Published in the United States of America in 1972
by Praeger Publishers, Inc.

Library of Congress Catalog Card Number: 71-161909

Printed in the United States of America

PREFACE

From January, 1969, through May, 1970, The Program of Policy Studies in Science and Technology at The George Washington University conducted a series of seminars on the processes and mechanisms of technology assessment under the direction of Louis H. Mayo and Raphael G. Kasper. The series was instituted to provide an opportunity for a relatively informal discussion of certain aspects of technology assessment. Participants were invited from academic, industrial, and governmental organizations in an attempt to gather as broad a spectrum of ideas and opinions as possible. At each session the author of a prepared paper moderated about two hours of discussion among the participants. This volume includes the ten papers presented during the series as well as summaries of the major points raised in the discussion sections.

The Program of Policy Studies in Science and Technology was established in its present form in 1966 under NASA Research Grant NGR 09-010-030. The Program has as its primary objective the development of a policy research capability within The George Washington University. The activities of the Program in the implementation of this objective point to the increasing attention being given to interdisciplinary studies, to problem-oriented research, and to the development of interinstitutional arrangements for analyzing social problem contexts, evaluating existing programs, and assessing the social impacts of prospective programs. As an integral part of the University, the Program is undertaking to develop skills in problem-oriented methodology, particularly with respect to problem contexts involving the impact of advancing technology on the processes, values, and institutions of society. It also provides a neutral forum for the discussion of policy issues by representative community groups and an advisory and research resource for policy makers in both the public and private sectors.

CONTENTS

Technology Assessment

CHAPTER

1

INTRODUCTION

Raphael G. Kasper

The rapid pace of technological progress has been accompanied by an increasing attempt to foresee and evaluate the consequences of technological activities. On the one hand, it is not hard to recognize that technology has yielded great benefits. However, on the other hand, it has become evident that not all of the consequences of technological innovation are beneficial and that some lead to quite serious, and often unexpected, hazards. The concept of "technology assessment" is developed in an attempt to understand and appraise the results of technology in order to promote the development of policies for the rational application of technology.

"Technology assessment" is a phrase that was introduced several years ago by Congressman Emilio Q. Daddario. However, it would be misleading to think, due to the newness of the phrase, that technology assessment is a modern concept. It is obvious that assessments have been made throughout history. What has changed, however, is the breadth of issues that are considered in the determination and evaluation of the effects of technology. Until recently, assessments were based on narrow considerations of technical efficacy and direct, observable results. But our growing awareness of the pervasiveness of the impacts of technology has led to the inclusion in the assessment process of the indirect, social and environmental effects that often accompany new technology. No longer is it sufficient to demonstrate that a new technique or product will result in greater speed, power, or efficiency, for recent history is replete with examples of technological advances that produce these effects only while they degrade the environment or dehumanize people, or threaten the existence of mankind.

Because of the popular concern about the environment and the highly visible character of some of the abuses that can be attributed to technology, technology assessment has been gaining a reputation as an essentially negative endeavor. Maybe this is inevitable; the risks are often so grave as to require great emphasis. But technology assessment need not be wholly negative. It may be that technology, if wisely applied, has the potential to solve more problems than it will create and to provide man with a fuller and richer life, free from the struggle for subsistence that has characterized earlier generations. That men might be moving from a view that technology is infallibly good to one that looks warily upon potential detrimental effects is perhaps because missed opportunities resulting from a failure to apply a particular technology are not as visible as environmental degradation that results from the unchecked progress of other technologies. It must be recognized that consideration of only the potential benefits of technology can lead to irreparable harm. However, at the same time, it must also be noted that consideration of only the potential risks of technology can lead to technological stagnation, with a concomitant loss of opportunity for the improvement of man's life. To be worthwhile, technology assessment must have a dual goal: it should seek to identify the ways in which technology may be utilized to fulfill the needs and desires of man while it seeks to minimize or eliminate the dangers of technological applications.

"Technology assessment," in my opinion, is an unfortunate phrase for describing the attempt to understand the consequences of the use of technology. There is a natural tendency in performing technology assessments to assume that technology can be considered as both a cause and a potential solution for all problems. This assumption may be true with many problems; but a mentality or a technique that seeks technological demons and technological miracles may be unable to identify nontechnological causes and cures. Although technology assessment, as was mentioned earlier, should examine broad social effects, the very name leads one to focus on a narrow range of considerations. If performed without care, it leads to the proposing of gadgets, of "technological fixes," to cure all problems, whether they be amenable to technological solutions or not. Still, the name apparently is here to stay; we can only hope that those who perform assessments do not allow the name to restrict their range of considerations.

This collection of papers and discussions illustrates clearly that there are no consensuses on many aspects of technology assessment; there is not even agreement on a definition. Many issues are raised and discussed in these pages: how assessments should be done, who should do them, how much information is necessary, how much information is available, and myriad others. Nevertheless, if there is one central issue, it is that of identifying the proper role of the expert and the citizen in determining the course of our increasingly complex and technological society. The questions raised outnumber the definitive answers: Can technological progress be explained to the public? What measures can be taken by interested laymen to make their point of view known in the face of the great accumulation of data and resources on the part of private interests? How can a true public interest affect the decision-making centers of government? Who can fund and who will listen to advocates for the public interest?

What follows is not intended as a final and completed study of the function of technology assessment, but rather as a presentation of views and ideas that will stimulate thought and, perhaps, action.

CHAPTER

2

ASSESSMENT INFORMATION SYSTEMS

Clarence H. Danhof

Long before man had begun to make purposeful use of fire, he, no doubt, had accumulated a stock of transmittable information comprehending effective and ineffective, and safe and hazardous materials and methods that related to his interests in accomplishing an objective, such as securing food. The knowledge of materials and methods that yielded desired results constituted the applied technology and was maintained since it was in constant use. Experience with materials or methods yielding results of neither positive nor negative significance no doubt tended to be forgotten. But special efforts were made to retain knowledge of materials or methods that proved so harmful in the results of use as not to be repeated voluntarily. The taboo served to communicate and preserve such knowledge.

The first action credited to man in the Book of Genesis is the naming of all the animals and birds--the foundation of an information system. Of greater interest is the third action there recorded, which was an effort to extend the available knowledge of resources with regard to their suitability for some desired use, in this case, food. The problem was the relatively simple and straightforward one of determining whether the fruit of a certain tree would, when eaten, provide satisfaction and yield undesirable consequences. The existence of a taboo against the use of the fruit of the tree suggested that earlier experience had been unfavorable. The possibility of undesired results

*Presented January 23, 1969.

was therefore of particular importance and invited consideration of whatever advice might be found. Adventuresome Eve sought the advice of the serpent who offered himself as an expert. The serpent assured Eve that "eating the fruit of the tree which is in the midst of the garden" would not produce death, but instead "your eyes shall be opened and ye shall be as gods, knowing good and evil."[1]

The advice was of compelling appeal. Eve did violate the taboo and ate and called upon Adam to confirm her satisfied reaction. Three consequences are recorded:

1. The immediate result was desirable: the fruit was good.

2. Adam and Eve became aware of their nakedness and invented clothes, an intermediate result subject to a variety of explanations.

3. Adam and Eve did encounter evil: they lost Eden. A technological system together with its related culture disappeared to be replaced by another.

In modern parlance, Eve was engaged in the assessment of new technology, a process which man has undertaken countless times over in his history. Though Eve's assessment effort was, viewed superficially, a simple operation, it had all the elements common to the modern process:

1. Initiative in identifying a solution to a felt problem or an opportunity to gain a desirable objective, both of which require exploration of an area involving some unknowns

2. The application of expert, specialized knowledge to the problem at issue, so that possible gains and hazards can be defined as clearly as possible

3. The possibilities that a new technology may yield desired or undesired results or, frequently, a mixture of both. The undesired consequences may affect the immediate user, a larger group, or all of mankind. Such undesired consequences may appear immediately, in which case cause and effect relationships are relatively easily identified; may emerge slowly, perhaps within the memory of a generation or two; or may require so prolonged a period of time as to be perceptible only in long retrospect. In the latter two cases, cause and effect relationships can usually be ascertained only by very advanced analytical techniques, if at all.

THE ASSESSMENT PROCESS IN CONTEMPORARY SOCIETY

The development of a new technology is, in the modern situation, usually a far more complex process than the determination of the edibility of a fruit. The development process is part of an elaborate effort to search out new knowledge and to appraise such knowledge for possible application to desirable objectives and the avoidance of those that are undesirable. Although the provisions made by society for dealing with these hazards has grown substantially, there is increasing evidence that the machinery is far from what it might or should be.

The assessment of new technology is an integral part of the process of accomplishing technological change. For our purposes the development of a technology may be viewed as occurring in two distinct stages, in each of which the assessment process differs in significant ways. In the preintroduction stage of development, it is the developer who, seeking to find a desirable use for a technique or material, applies tests he deems relevant to determine both the positive and negative characteristics. In the application stage, it is the user who applies such tests, employing what may be somewhat different criteria. An essential part of the assessment process, then, involves the using public which provides a large part of the effort in seeking those conditions under which the benefits of the new technology can be best attained. That such efforts continue over a prolonged period of time, permits the longer-term effects of the new technology to make themselves apparent.

The decision by the developer to make the innovation available to the public through the market or otherwise is presumably made with some confidence that it will perform a desired service at an acceptable cost-benefit ratio and that it will do so without undesired consequences if prescribed conditions of use are followed. In the usually long process of developing an innovation, numerous actions have occurred that yielded negative or undesired results. Both contribute to the originator's knowledge of conditions under which the innovation must be applied to secure hoped-for results.

The decision to introduce may have rested wholly on the basis of information developed and assessed within the originating organization. In many cases, however, outside assistance has been sought. The device may then have been user- (market-) tested. It may have been submitted to examination by experts for conformity to standards established by an industry-wide organization, by an independent testing organization, or by a government agency. In the case of new drugs and certain food or food-related items, it will have been clinically tested and then submitted to the Food and Drug Administration for its approval. When finally offered to the public, the considerations favoring public experimentation with the innovation are substantial. No unfavorable results accompanying use have been identified, or if some have been, it is believed that they can be eliminated or held within acceptable levels by advice and warnings controlling the circumstances of application. The first stage of assessment has been completed.

The central characteristic of the second stage of assessment is the involvement of the general public. No testing under limited or even simulated conditions can completely comprehend all the possible circumstances of use. Public usage is likely to present a more comprehensive range of possible applications under widely varying circumstances and hence to yield a wider variety of results, both positive and negative. Some undesirable effects will emerge under unanticipated conditions and only over a more or less prolonged period of time. The using public can be expected to communicate its opinions of the innovation to the originator in numerous ways. The rates of purchase will indicate acceptance with more or less unqualified enthusiasm as may also the appearance of competitive suppliers. Potential users may, on the other hand, cease to purchase or return the device if warranty arrangements had been made.

We are concerned here, however, not with performance that falls short of expectations nor with adverse effects that are merely unpleasant to individual users. Our concern is rather with adverse effects of the new technology and the relationships that follow between users and originators. Such adverse effects may result from the failure of the originator to provide proper materials and instructions or to anticipate results that could reasonably have been forecast. Undesirable results may also follow because of actions by users beyond the reasonable anticipation of the originator or sponsor. It is true that new technologies are sometimes viewed and accepted with great caution, particularly when they constitute new complex systems that make fixed capital investments obsolescent and require substantial new capital investments. Other new technologies are quickly accepted and may then be applied less critically and with less discrimination than occurred in the late stages of the originator's testing. Such was the case with, among many others, penicillin and other early types of antibiotics, some medical techniques such as x rays, ovarectomies, appendectomies, and tonsillectomies, and such pesticides as DDT. While reports of undesirable effects of application accumulate in the professional literature, are summarized and evaluated, constraints may develop. This occurred in the past, for example, with some uses of arsenic and lead and seems now to be occurring with such materials as mercury compounds and hormones. The process of reevaluation may be effected solely by users and the sponsor, by members of the expert group involved, or also by nonexperts, as in the current controversy over the use of DDT.

In this second stage, the problems of observing, recording, analyzing, and communicating the results of application may be more complex than in the preintroduction stage since more or less numerous appliers are involved and the circumstances of application may vary widely from those of the first testing stage. The probability of unanticipated adverse effects therefore increases. At the same time, the observing, recording, and analyzing of such adverse effects, occurring within the originating group in the first stage, becomes difficult because of the numerous, scattered participants and the need to establish cause and effect relationships in what are frequently exceedingly complex situations that operate over long periods of time.

One broad reaction by the society is to require the sponsors of new technology to undertake more elaborate preintroduction tests. Such efforts are strongly constrained in many cases by consumer interest in enjoying the apparent benefits, by the possibility of the existence of a "moral issue" in the witholding of a new technology.[2] and by great reluctance to discourage the innovative process as by increasing its costs. In a society in which growth is a basic desiderata and in which it is assumed that problems can be solved by technological discovery, the witholding of a new technology, even for a limited time, tends to be viewed as an offense to society.

ADVERSE EFFECTS: DEATH

The most earnestly avoided adverse effect of new technology is human death. Death is also one of the most carefully recorded events in our society. Since protection against controllable or avoidable death is a legal right, the cause of each death is determined and reported to responsible authorities. Assurance that such recording is made follows from the fact that burial, in most jurisdictions, requires a permit which in turn requires that certification as to the cause of death has been made by a prescribed official, an attending physician, a public medical officer, or in their absence, a coroner, who may call upon experts and upon a jury to assist him.

Causes of death are reported by use of a standardized terminology that may reveal little or nothing about the underlying or contributing developments that were involved. In some cases involving new or relatively new technology, cause and effect relationships are easily determined by the expert and perhaps by the untrained. In many others, the large number of factors that may be suspected present very difficult analytical problems. The investigation of possible causal relationships between technology and death is, therefore, not a routine operation, but one that rests upon professional initiative in undertaking explorations of the factors that might have been of causal significance. Repeated observations of experts reported in the professional literature may lead to statistical correlations that suggest causal relationships. Reports of such statistical inferences may then suggest hypotheses stimulating research into the processes suspected.[3] Resting upon an accumulation of experience with frequently subtle relationships in areas characterized by the presence of numerous variables related in complex ways, the process is frequently time-consuming, as in the case of the relationship between cigarette smoking and emphysema.

ADVERSE EFFECTS: PERSONAL INJURY

Many new technological innovations present few, if any, intrinsic dangers. Many others can be adapted to include built-in safety provisions or require only simple instructions and cautions that, it is assumed, will assure safe usage, given a public of reasonably high levels of functional literacy. The benefits of other innovations are obtained indirectly, in that the public secures the services of the technology through an expert and does not itself control its application.

A death attributable to the use of new application of technology represents a failure of the efforts of the originator, and possibly of the larger society, to anticipate hazards, to impress on the user the existence of a recognized danger, or to provide methods whereby the desired services can be obtained safely. It is an inherent part of the process of developing a new technology that hazards be identified and that the procedures involving hazards be eliminated or minimized. The originator will have acquired or have available a stock of information on hazards and their avoidance. In some areas, the originator--and the public--have available testing services that provide constraints by the application of standards developed from experience. These form the basis for the formulation of rules for safe application, of which the user should be fully informed. It remains true, however, that in public usage the circumstances of application will be more varied, and on occasion less guarded, than they were in the originator's laboratories. If a hazard exists, sooner or later some member of the public will experience it and the critical question is that of frequency. Data on injuries by type of technology are by no means as accurate as death data, although causal relationships are in many areas more readily established. The collection of data on public experience is then an essential function; it contributes to the continuing development of the technology and to the procedures governing its safe use. One of its products is the development of data on probabilities of adverse reactions, these being essential to the determination of the risks being accepted in the usage of any given application of technology.

Such analysis of experience has led to a variety of public actions to assure safety. These may be categorized as follows:

1. Legal prohibition of the use of a technology, e.g., opium, LSD, lead for water piping, x rays for shoe-fitting (Note the failure of the effort to prohibit the beverage uses of alcohol.)

2. Legal prescriptions of the conditions under which a technology may be used, e.g., prescription drugs, electrical and plumbing installations, nuclear energy, x rays, aircraft, vessels. Generally, such regulation utilizes licensing, certification, and inspection procedures, and typically is applied to situations in which benefits of the technology are provided to the public through an intermediate supplier of the service.

3. Legal requirements that the user be informed of risks and of limitations on use procedures, e.g., pesticides, cigarettes

4. Legal requirements that a given technology must be used as part of a system, e.g., aircraft, certain regulations pertaining to air and water pollution, federal auto safety legislation (Note that licensing procedures cited under 2 above may include such positive requirements.)

ADVERSE EFFECTS: COLLECTIVE INJURY

In the preceding discussion, attention centered upon injury to a person or relatively small number of persons occurring within a relatively short period of time as a result of some direct contact with a technology. Adverse effects can also be experienced by persons--possibly very many--because the environment within which they live includes adverse forces that may accumulate into uncontrollable and eventually disaster proportions. Many problems of this nature that have arisen in the past have been successfully dealt with, as in the development of public waste disposal systems or in the substitution of processed and chlorinated water for water naturally "pure." Indeed our society rests upon past successes of this nature.

Such "fixes" will no doubt continue to be effected, but there is much evidence that they will become more difficult to obtain. The possibility also exists that both because of the nature of a solution to a certain problem and because of the unavailability of a technological fix, society may find itself forced to change its structure and values to deal with some adverse effects of new technology; the alternative might be disaster.

It is the long-range effect that presents particular problems, partly because the difficulties of determining the causes of adverse effects are numerous and partly because the values by which adverse effects are judged are not a matter on which consensus is easily obtained. Many a new technology has made possible the achievement of an immediate goal only to produce over a period of time new problems of great difficulty. Viewed broadly, many of the problems arise from the very success of man's efforts to expand the very technology that has created problems that threaten his achievements if not the existence of society itself as now structured. Excessive population growth rates is one of the more obvious aspects of the problem.

The experience of Adam and Eve is again relevant. That they were correct in their assessment of a new technology--up to a point--is obvious. Man has subsequently greatly elaborated upon their procedures. The disaster that befell Adam and Eve we may wish to consign to myth or fable, or to the wrath of God. If it were a unique experience, we might be justified in the view that it holds no significance for us. It was, however, not unique but rather, as Arnold Toynbee recounts in some detail,[4] a frequent experience over human history. We know little about such societal catastrophes. Numerous forces were no doubt at play. That technological failure was a component of many, if not most, seems a reasonable hypothesis.

As man's mastery of his evnironment has expanded--enormously in the Western World of the twentieth century--his relationship to that environment has become increasingly brittle and fragile. Man's facile assumption that he can manipulate his environment to his own exploitative objectives without otherwise changing that environment in some basic and perhaps critical characteristics is an assumption that appears increasingly dubious and perhaps perilous. Man's ecology clearly has changed, and we are increasingly aware that not all the changes are improvements. If, then, man increasingly cannot assume his "natural" environment as a given, because he has changed it, enormously important questions arise. It is the nature of the man-made environment that requires analysis and careful monitoring.

In this category of assessment problems, the originator of the technology has passed from the scene as a central figure since technology is now thoroughly accepted and highly valued for its benefits. The users and beneficiaries of the technology now constitute a vested interest. The technology is no longer new; it is thoroughly embedded as a component of the society.

The problems presented by the technology are dealt with by procedures that differ sufficiently from systems previously discussed to warrant designation as a third type or stage of assessment. It is in these areas that the role of the expert and particularly his interaction with the public is least developed and most uncertain.

INFORMATION SYSTEMS AND THE EXPERT

The preceding effort to define the range of assessment problems gives emphasis to what is undoubtedly obvious. Since the days of Eve, man has relied upon expert observation and analysis as the foundation of its efforts to deal with technological change. As a result of the increasing probing of knowledge for its application potentials, the expert has become both quantitatively and qualitatively more important than in Eve's day. The experts ferret out potentially useful technology, are principal source of advice and persuasion, and by serving as gatherers, synthesizers, and interpreters of observational data and analysts of causal relationships, are a principal source of information regarding the results of usage.

Society is dependent upon its stock of qualified experts for the following:

1. To ask relevant questions and record their observations

2. To communicate with each other to confirm their observations and to establish magnitudes (It is here that the growth of knowledge has produced numerous problems of information organization and retrieval. These problems, only as they affect the function of the expert, relate to the assessment process.[5])

3. To summarize their findings, preferably as a consensus,

but in the absence of a consensus, then as to areas of agreement and disagreement

4. To indicate the significance of these findings to an audience capable of determining that action is or is not warranted.

On its part, society must provide such an audience, serving as a source of guidance as to areas of concern, to identify areas of agreement and disagreement and, most of all, to assert the nature of the values and objectives involved.[6]

In the case of Eve, the expert did not hesitate to communicate enthusiastically the advantages his advice offered his client. In this respect his performance differed little from that to be found in much contemporary popular scientific literature.[7] The experts did include some predictions about the long-range consequences that might follow and their effect on man's society. As experts do, the serpent left to his client the choice of what long-range values she wished to pursue, and the evaluation of the gains that might be made as compared with the risks she was prepared to assume. Then as now, expertise did not carry with it any special qualifications regarding values one wishes to pursue.[8]

As the expert becomes more important, societies develop methods of dealing with him. A first step is to establish and protect confidence by requiring some procedure to distinguish between the qualified expert and the charlatan. In both the routine exercise of the expertise and in the discovery of new knowledge, a variety of checks exist, although all of them tend to rely on the peer system, whereby experts certify the competence and the quality of work of others in their field. A matter not easily determined is the extent of confidence an expert group merits, a question that from time to time requires nonexpert examination of the assumptions and methods upon which the expert's claimed knowledge is based.[9]

Far more difficult is the question of the use that the nonexpert public should make of the expert. It is a well-established principle that the "expert be on tap but not on top," at least in policy-making.[10] In form, that principle is well-observed; in practice, there are difficulties. The public cannot expect to obtain from the expert what it feels it needs unless it is able to ask questions that will elicit appropriate answers. Recall that Eve's source of expert opinion is referred to, after the fact, as a serpent. Considering the consequences to her of the expert's advice, Eve's opinion was no doubt justified.[11] It is at least possible that the relationship might have been more satisfactory

Eve asked her expert some questions that pushed the analysis into a more critical consideration of long-range consequences. In the area of assessment, as in others, effective relationships between the experts and the public rest upon interactions in which each urges upon the other consideration of its concepts of problems and values. Flexibility in this respect seems to require an avoidance of "officially" sanctioned expertise and other forms of rigidity that hinder the flow of observations and analyses and that may lead too quickly to consensuses in which values and underlying assumptions remain unquestioned.[12]

The nature of the problems differ with social experience in the area of the technology involved. Where legal responsibilities have been assigned, reporting procedures develop into well-structured systems, as in the case of human fatalities. This is not to say that the operation of that system might not be improved, but that the weaknesses that appear to exist lie with the discretion exercised by the expert observers. We can reasonably doubt that the cause of all deaths is determined with any high degree of accuracy. However, from the point of view of assessing the technological source of fatalities, it seems likely that over time significant experience will be identified and reported with reasonable promptness. In any case, very considerable progress has been made.

The reporting of technology-related injuries is less satisfactory, with the exception of those areas for which specific responsibilities have been legally assigned. In the case of aircraft fatalities and pharmacological and industrial accidents, legal requirements and compensation provisions seem to assure substantially accurate reporting and to stimulate efforts to reduce hazards. In other areas, the reporting procedures are less satisfactory, data collection is frequently haphazard, and legal and financial liabilities are uncertain, with the role of human negligence a source of confusion. Corrective measures are usually of the advisory and educational types.

In areas involving collective safety, such as water and air pollution, significant steps towards assignment of legal responsibility recently have been taken. In both cases the evidence has been sufficiently persuasive to support such action, but much data collecting remains to be done before the precise nature of the problems are known or corrective procedures are developed. Meanwhile, data collecting is enhanced through the enforcement machinery provided. Establishment of standards provides a guide for observation and measurement as well as a focus for debate over the values involved.

The pattern by which human waste pollutions are controlled

might be applied to other pollutants, such as air pollution, solid waste, noise, heat, and debris in scenic areas. Environmental control is a concept of some appeal, but the wide variety of the problems and values it comprehends suggest that the approach will be component by component. These and related problems such as the population problem, are in the stage in which information is developed and exchanged principally within the research community and with those members of the public already convinced of the seriousness of the problems. Some members of these groups are carrying out the functions of synthesizers and publicists and some are organized for the purpose of influencing legislation.

In a mature communication structure, the process of observing, collecting, and disseminating information will reflect the interests and values of the responsible organization. Even when a formal organization exists with responsibility to act upon technology-generated problems, informal observations and reporting continue as an important input to the assessment process. The role of private groups as sources of information and appraisal is particularly important when there are differences in the values held by the responsible agency and those held by private groups. In the case of DDT, for example, the responsible government agency has contented itself with controlling adverse effects of usage by requiring cautionary labels. Some private groups believe that sufficient evidence has been accumulated to demonstrate clearly that the continuing use of the material is too dangerous to nonhuman life, and eventually to human life, to countenance. A principal problem of such groups is to find an audience that can react to its findings in such manner as to contribute to the formulation of public policy. In this instance, the private groups are bypassing the relevant agencies of the Federal Government. They are presenting their case instead to state regulatory agencies and may eventually appeal to that fundamental structure, the court system.

If, as in the case of DDT, the experts disagree about the scientifically determined facts and the economic values involved, the public finds itself in the position of determining the basis for the disagreement. Frequently the disagreement rests upon inadequate knowledge, and the solution is a continuation of research on the problem. The public then must recognize that the best available scientific knowledge at any given point of time is subject to change, and sometimes to very drastic change. Permissible levels of exposure--of radiation or pesticides--are, for example, frequently judgments by experts rather than well-documented determinations. We must expect these to change over time, though the direction of change is difficult to predict. In such common situations, disagreements among experts stem from interpretations made about the significance of available knowledge; and the crux of an assessment is in the application of values frequently held implicitly rather than explicitly. In the case of DDT and in many others, there exists a conflict of values. The expert cannot be assumed to have any special competence in matters of values that a public group or society as a whole may choose. The public cannot and should not passively accept the values of the expert as its own.

CONCLUSION

One can argue that the apparent evolution of assessment procedures is proceeding along lines that will provide adequate control machinery when the need is clearly demonstrated. This may mean a continuation of our traditional approach of responding to a present or immanent crisis. The nation has done better than that in some cases, but only time will tell if we have been properly alert in others.

If the earlier comment regarding the fragility of man's control over his environment is sound, then our society faces problems of an urgency not easily measured but assuredly reflecting the very rapid rate of scientific and technological change. We may, in fact, face the prospect of irreparable crisis. Even the possibility of such a crisis suggests that reliance upon the slow evolution of public acceptance of responsibility is unsafe.

In this situation, it should be recognized how great is the difficulty facing expert observers in those areas that lack a formal information structure, particularly those that are viewed in a long time perspective. The problem is not that there is a total lack of expert warnings of possible undesirable developments. Our society, like most others, has a fair number of prognosticators who serve as Jeremiahs. The difficulty is the absence of a platform and an audience before whom their assessments can be given adequate and authoritative consideration in the light of long-range public policy. The President's Scientific Advisory Commission, the National Academy of Engineering, and similar organizations have acquired important though limited functions in providing authoritative reviews of expert thought.

In the last analysis, the problem is one of identifying the social significance in an assessment situation. At this stage the assessment is no longer solely a function of the expert, but it involves the values and objectives of the society and frequently the conflicting values and objectives of groups within the society. The problem must then be resolved through the political structure of society. In the political process the role of the expert as advocate is no doubt necessary since he is most knowledgeable and most involved. As advocate, however, he presents special problems of credibility to the interested public, particularly since he frequently if not normally takes positions well beyond what can be documented by information generally accepted as true. Securing public acceptance of an action affecting its values and objectives is not difficult if there is a clear threat to life. In situations where the current threat is a low statistical probability or if the threat is remote in time, securing public action is far more difficult. It is with regard to such problems that the nation's information system is precariously ineffective. One approach to such problems might be a periodic, authoritative, public review of the nation's progress toward its scientific and technical goals, among which is the protection of its hard-won achievements.

NOTES

1. Genesis 3: 4-5.

2. This occurred in the case of the Salk vaccine. See U.S. House, Committee on Science and Astronautics, Technical Information for Congress, a report by the Science Policy Research Division of the Legislative Reference Service, 91st Cong., 1st sess. (1969), p. 321.

3. It is not suggested that the system is beyond criticism. For extensive criticism with reference to the evaluation of new drugs see U.S. Senate, Committee on Government Operations, Interagency Drug Coordination, 89th Cong., 2nd sess. (1966), particularly pp. 102-262.

4. Arnold J. Toynbee, A Study of History (6 vols.; New York: Oxford University Press, 1934). Unfortunately Toynbee's analysis pays scant attention to technological factors.

5. These problems are surveyed and assessed in National Academy of Sciences, National Academy of Engineering, Scientific and Technical Communication (1969).

6. Philip Ritterbush, "Environment and Historical Paradox," General Systems, Yearbook of the Society for General Systems Research (Washington, D.C.: Society for General Systems Research, 1968), XIII, p. 112: writes: "The problems in environmental improvement at the Federal Government level should not be characterized as deriving from too little scientific knowledge, but from the failure to devise a suitable forum within which conflicting expert views could

be reconciled and comprehensive program planning could be carried on. . . ."

7. At the present time, the reporting of the possible results of new and hoped for knowledge in genetics is markedly characterized by prophecy. See, e.g., Albert Rosenfeld, The Second Genesis: The Coming Control of Life (Englewood Cliffs, N.J.: Prentice-Hall, 1969).

8. A still useful analysis is Harold J. Laski, "The Limitations of the Expert," Harpers, CLX (December, 1950), 101-10.

9. Joseph Jastrow, ed., The Story of Human Error (New York: 1936) makes for disquieting reading in this respect.

10. The aphorism is attributed to Harold Laski by Paul H. Appleby, "Making Sense Out of Things in General," Public Administration Review, XXII, Appleton Century (December, 1962), 175.

11. Note too that the serpent was severely punished by higher authority. Genesis 3: 14-15.

12. For one of many such, see Ashley Schiff, Fire and Water: Scientific Heresy in the Forest Service (Cambridge, Mass.: Harvard University Press, 1962).

DISCUSSION OF ASSESSMENT INFORMATION SYSTEMS

The discussion centered on three general topics (the role of experts, the adequacy and destination of information, and the role of the public) and touched briefly on several others. As might have been expected, there was little consensus: a broad range of opinions was presented by the twenty-six discussants.

ROLE OF THE EXPERT

The seminar participants differed on the extent to which reliance should be placed upon "experts" in obtaining the information necessary to perform assessments. The views ranged from those of a Congressional staff member who believed that only "specialists and experts" could solve the problems of technology assessment to those who indicated some reservations concerning the great dependence upon experts that seems to characterize assessments. A university professor was concerned with the fact that experts can, and often do, differ on matters of substance. "I'm sure," he stated, "that we've all seen a number of cases in which both sides have witnesses with expert credentials and yet come up with completely opposed views." A Government official raised the problem of "finding unbiased experts. It seems to me that the experts who are most capable of making technological assessments are liable to be in the employ of bureaus or companies that are interested in furthering technology." This developmental bias of experts often makes it difficult to find technically competent opponents of a particular technological application. In the words of another university professor:

> I suspect, today, that there are at least several areas of technology which are controversial, at least to the layman, with respect to which it is almost impossible to find an expert who will take the anti-technological point of view so that the negative view never gets articulated, whether it's sound or not.

One suggestion for overcoming the "biased expert" problem was provided by an agency representative who held that greater use of university scientists in assessment might overcome the predominant influence of development-minded industrial scientists. The problem of locating impartial experts was, in his words,

> . . . one of the best arguments that I know of for us having a broad scale federal support of research in the universities and gathering up a lot of these functions (of obtaining information for assessment processes) within the free play of the intellectual marketplace in the academic sphere. There you do have the whole tradition of public disclosure and public discussion--in fact, the whole system works on the public nature of the exchange.

The universities are, in part, shielded from those market pressures that make it difficult, if not impossible, for industries or agencies to undertake candid and impartial assessments.

Some of the participants, including a university scientist, noted a serious communications problem that exists between technologists and nontechnologists. There is a need, in the view of an agency official, to "vulgarize" technology ("and that meant in the best sense") in order to make it understandable, "if not to the man in the street, at least to the average Congressman." The use of the mass media was suggested as a means of increasing communications between the scientific and lay communities, though it was pointed out that journalists possess a vested interest in their quest for higher circulation, which often leads to inaccuracy in the pursuit of sensation.

Part of the concern with the reliance on experts was based on the contention, presented by a Government agency representative, that the experimental skills of scientists and engineers do not necessarily coincide with those skills necessary for identifying and analyzing the consequences of new or existing technologies. A university scientist disagreed and contended that, in fact, the experimental research outlook may give the best possible information for assessment purposes. Although one of those present attempted to minimize the problem of just who the experts are by holding that experts only make suggestions and recommendations and do not perform the actual assessment, another took issue with this point:

> I think that to the extent that we have technology assessment today, the assessment is made by experts. To the best of my knowledge there is no agency of the U.S. Government today that assesses technology that does not have, on its staff a significant proportion of technologists or scientists.

While suggesting that experts "may have some competence to evaluate the risks" of a technology, the speaker, a university professor, doubted that technological experts have the competence "to determine, let alone qualify, the benefits." Perhaps, he added, a wider membership in the assessing group, including social scientists and lawyers as well as scientists and engineers, might lead to a more effective mechanism of assessment. "The process we are talking about," he concluded, "needs larger doses of common sense and smaller doses of expertise."

ADEQUACY AND DESTINATION OF INFORMATION

Several participants questioned the adequacy of existing information about the consequences of technologies. One, a university researcher, believed that the prepared paper had been "overly optimistic" in its implication, that required information is either in hand or relatively easy to acquire. For both existing and ongoing technologies, current sources of information are, in his opinion, not sufficient and "there are serious conceptual problems in many cases as to how you would actually develop the information. I think in a lot of cases you would have to undertake extensive research programs to develop the kind of information you needed." He expressed some concern with the fact that in most instances, we lack institutions with the resources, responsibility, or incentive to generate the necessary information. An agency representative saw a need for greater specification and accuracy in the reporting of consequences of technological applications; a lack of such detailed reporting has hindered assessment processes greatly. It was noted that virtually all police reports of automobile accidents, for example, cite as the cause "excessive speed and drinking" because attributing an accident to such a cause is easier than performing detailed and complex studies to determine the actual cause. That this kind of reporting, done for purposes of expediency rather than accuracy is prevalent, leads to the conclusion that statistics are not always a very meaningful source of information about the effects of technology.

However, if information systems are not satisfactory for existing technologies, what then can be said of the performance of such systems in identifying the effects of new or developing technologies? This question was raised by a Government official, but aside from noting the complexity of any effort to foresee future consequences, the participants declined to elaborate further and limited the remainder of the discussion to the "postintroduction" phase of technologies.

Several speakers raised the problem of determining what occurs within private industrial firms. This is quite an important matter since, as was noted earlier, qualified experts in any given technological field are almost always employed by industrial concerns which are involved in the development of the technology. A Congressional Aide noted the "tremendous difficulty of obtaining the highly essential information that is in the files of the large corporations," while he emphasized that "we've got to get very honest reporting from the private sector in order to know exactly what's going on in terms of possible dangers." Some new Senate bills, he indicated, may have the effect of opening up for public consideration some results of individual research.

The destination of any gathered information may play a large role in determining the type of information which is to be obtained. To one university professor, this is the "crucial question." After all, he claimed, when no mechanism or organization requires information, when can information be termed inadequate; "as long as nobody wants the information, in any detail except for statisticians, any information is sufficient," regardless of how sparse it may be. It is only when an organization exists to examine the information that one can begin to be concerned with the adequacy of the information. The need for an opportunity to present information, to have a hearing, was repeated by several participants in the course of the session. "Whether or not the issues were accurate was not the point," stated one speaker in describing his personal difficulty in obtaining a hearing. "The point really was whether there really was an opportunity to discuss them." Often, those who seek to obtain a hearing, he claimed, are resented and refused any opportunity to testify.

A mild dissent concerning the emphasis on information systems was presented by a Government official. He made it quite clear that he believed that information, or even information systems, could not be considered in isolation; "When I consider the quest for information systems to be used in this context, I have to ask myself: information systems to what ultimate end?" He went on to decry what he termed a cult, spawned by improved data collecting capability, which seems to hold that better information handling alone is a solution to the technology assessment problem. This is a false view, he claimed:

> You could go on accumulating mountains of information about technological problems that go on and won't quit. The issue to me is--what are you going to do with it? Who, in the end, is going to organize the political power to translate the information into corrective action. I'm more interested in that end of the problem than I am, frankly, in accepting the presumption behind the argument for information systems because I think we can exhaust a great deal of energy and resources with the best of intentions and yield very little [if we concentrate on information systems alone].

The point, he said, is that expert opinion, and even common sense, haven't produced effective action. (It may be noted that chapters 3, 4 and 5 are concerned with the use of information and its translation into action.)

PUBLIC PARTICIPATION

The discussion turned to the question of public participation in the assessment process. In commenting on the prepared paper, an agency representative noted that it

> . . . says very little at all about the extent of public participation in the technology assessment process. In our democratic society, the way we generate political pressure is to have the public involved and to have some real broad scale public interest. What we don't really seem to

> have is a way to match the expertise of our expert to the intellectual resources of the public. If we concentrate on something like a technology assessment agency that does not allow for public advocacy, we really won't have gotten anywhere at all.

It was his view that technology assessment is performed almost secretly and outside the usual framework of the democratic process. The situation can be corrected: "I think that many of the things that we find to be very serious public problems that do not get adequately ventilated would be adequately ventilated if the recommendations of the scientific advisory apparatus were public and its documents freely available to the public." The importance of television as an educator of the public was strongly emphasized by a Congressional Aide.

To the suggestion that assessment proceedings be made public one participant expressed concern that "if the assessment process is too public, if all of the preliminary recommendations and suggestions and some of the wild ideas are publicized at the earliest stages, then you will deter the kind of candor and imagination you want to encourage" on the part of the assessors who might be leery of pursuing certain potentially fruitful lines of inquiry in the face of public disclosure of their efforts. He felt that it was necessary to strike a "fairly complicated balance . . . between making the final recommendations open and available for public scrutiny, which is not always done, and making every preliminary step in the assessment process available to the public."

Some of those present had reservations about the role of the public in assessment. An educator noted the characteristic tolerance of the American people toward technological change. "The man on the street," he stated, "will tolerate anything . . . in behalf of the advancement of technology." Unless some change in this prevalent attitude can be induced, the public will remain ineffective in producing meaningful action regardless of how well educated it is about the nature of the problem. But even if the public inertia could be overcome, the public role might not become an effective one, at least in the view of one speaker. He believed that public involvement is important only in extreme cases. Smaller issues cannot excite the public interest and the majority of assessment problems fall in this category rather than in a crisis class.

OTHER CONSIDERATIONS

The remainder of the discussion touched briefly on some miscellaneous points of interest to the participants. A Congressional Aide and a professor expressed some concern with the fact that technology assessment seems to concentrate on the detrimental aspects of technologies. They asked for more perspective in noting the positive aspects of technological development. Another Congressional Aide held that Congress does in fact, assess both benefits and risks and noted that "Senator Muskie and Congressman Daddario always use the word 'benefit' first and then they talk about hazards." Others felt that the problem of identifying benefits is not crucial to the assessment process. In the words of a university professor,

> In the very nature of things one can always assume that the benefits of technology will be more than adequately articulated because there are very powerful vested interests which want to articulate those benefits. The real problem is how you articulate the costs and the risks.

An agency official agreed and suggested that all new technological developments be subjected to energetic opposition for the purpose of exposing risks which are either unnoticed or neglected by proponents of the technology. (This argument is pursued further in Chapter 4, in which the central concern is the adversary process in technology assessment.) Perhaps the position of many of those present was summed up by a Congressional Aide who said of the consideration of risks and benefits that "the two go together--they always go together and they have to go together." Any assessment, to be complete, must consider both.

Two interesting and practical examples of the current state of information systems and assessment discussed deserve some note here. One of these concerned the setting, by the National Committee of Radiation Protection, of radiation exposure standards for workers in the atomic energy field. These workers are permitted to receive radiation doses ten times as large as those to which the general public may be exposed. The decision to set higher limits on allowable doses for industry personnel was arrived at without any consultation with the workers, and it was predicted by one seminar participant that this disregard for ordinary collective bargaining procedures might lead to strikes in the nuclear industry.

The other example was the establishment and operation of the National Transportation Safety Board, which is required by law to make its findings and recommendations available to industries and the public. The Board operates solely in the interest of the public and, in the view of one discussant, has been effective in inducing changes in the transportation industry through a combination of letters to offending industries and news releases that have notified and educated the public. There was some skepticism among others present over how the effectiveness of such an organization can be evaluated. However, it was suggested that despite the fact that the Board has no way to apply legal sanctions to any safety offender, the letters and publicity had produced some changes in conditions in the interests of increased safety.

CHAPTER

3

TECHNOLOGY ASSESSMENT AND THE CONGRESS

Richard A. Carpenter

This paper is designed to take advantage of the Seminar Series to elicit opinions from the participants on a particular matter: the Legislative Branch of the Government as the major focus for technology assessment. My thesis is that the Congress is the most important user of such information and that a bridge to assessment capabilities must be built. The details of its organization, activities, and relationships to other agencies and the private sector are not yet developed. The thoughts expressed herein must, of course, be considered as personal opinions and not charged to the Legislative Reference Service** or to any Member of Congress.

Having been involved in the development of technology assessment for some years, I am not hesitant about offering a definition, even though such an act entails risks of diversions into semantic nitpicking.

Technology assessment is the process of identifying and studying consequences of the applications of science. The objective is to

---

*Presented February 19, 1969.

**Since this paper was presented, the name of the Legislative Reference Service of the Library of Congress has been changed to the Congressional Research Service.

optimize the use of knowledge for the benefit of society, to enable us to manage the affairs of a complex, technology-based civilization. Consequences that are unintended, unanticipated, and unwanted are to be minimized. Assessment includes forecasting and prediction, retroactive evaluation, and current monitoring and analysis. Measurements involve noneconomic, subjective values as well as direct, tangible quantifications.

In the United States, the combination system of free enterprise democracy is the vehicle for technological change, economic growth, and social progress. The marketplace continually weighs and balances costs and benefits. The political process performs a similar function for governmental programs. Risk-taking is accepted in America as necessary for private profits and public welfare gains.

In the past decade, a dramatic challenge to this method of operation has been recognized. A much more sophisticated view of technological change is required. The close interdependence of mankind and the natural environment is increased by the swelling world population. The social environment is sensitized to untoward effects by the concentration of people into cities. Hidden economic factors often outweigh obvious market values. New forces, available from the application of science, have the power for worldwide, long-term, --and often irreversible--changes.

Thus, the development of a technology assessment capability is an urgent task, one recently imposed on our economic and political systems by the exponential growth of population and science.

The harbingers of technology assessment include nuclear weapon testing, persistent pesticides, nonbiodegradable detergents, smog, thalidomide, automation, the computer revolution, and so on. A hundred less spectacular events could be listed from many industrial and environmental situations. The rubric is broadening to include the often trivial concerns of consumer protection, the intricacies of a sea level Panama Canal, the worldwide effects of fossil fuel combustion and the cosmic concerns of planetary contamination. All these questions boil down to a need to know more.

I would add a special aspect to technology assessment, that is, the positive value of knowing more about our actions. The alarms and restrictions can too easily become the dominant theme. In fact, assessment can have great value to commerce and to government in revealing unknown benefits and providing confidence for decisions, which otherwise would not have existed. Falsely based inhibitions can be just as bad for society as undesirable consequences.

In summary, I see technology assessment as the final step in a long sequence that could be termed the socialization of science. In this sequence science has been apart from the mainstream of society for most of recorded history. Up until the end of the nineteenth century science had its own momentum and direction, asking virtually no support from its host culture and directly contributing little in return. Then, with the industrial revolution, agricultural engineering, and applied research in the early 1900's, and the military awakening to science in the two world wars, the relationship to society grew rapidly. Science promised and delivered. Society sponsored and glorified research without question.

This arrangement continued until the late 1950's, when the governmental and corporate support of science reached such financial proportions that a return-on-investment analysis was inevitable. Most manufacturing organizations had erected research laboratories--for window dressing if nothing else. The Federal budget included $4 billion for military and atomic energy R&D in 1958. When the space program (with its strong basis in the goal of international prestige) was added the R&D budget rose to $14 billion in 1966 for the three agencies of the Department of Defense (DOD), the Atomic Energy Commission (AEC), and the National Aeronautics and Space Administration (NASA). In fact 90 per cent of the total Government R&D support in the last decade was for these purposes. In this period, roughly 1956-66, the concept of efficiency in the application of science emerged.

The R&D practitioner was asked to carefully plan and execute his science in return for generous support from society. The systems approach to projects was perfected. Invention was put on schedule. Spin-off and fall-out benefits were extolled by NASA. The DOD procured R&D performance on incentive fee contracts. As additional agencies perceived the ways in which science could help accomplish their missions, the public needs of pollution, transportation, crime, education, housing, etc., began to be served. The demand that science be relevant was accepted. The ability of any scientist, particularly the applied technologist, to work outside this requirement diminished.

Now we are entering a final phase. It is no longer enough that applied science be relevant. So pervasive is science in this civilization that it is difficult to think of any irrelevant experimentation. But, through technology assessment, society is insisting on eating its cake and having it too. Only when the course of innovation has been studied and the benefits are shown to outweigh the costs, is technological innovation welcome. Each unwanted consequence must have its "technological fix," in Alvin Weinberg's term. Consumer protection replaces caveat emptor because it is beyond the ability of the buyer to beware. Technology assessment is an added responsibility of science, to be incorporated into the research-development-innovation-diffusion sequence at every stage. Assessment is not measured solely in technical parameters; but the measurement is a duty of the managers of science.

Thus in 1963, in his memorable address to the National Academy of Science, President Kennedy said, "Every time you scientists make a major invention, we politicians have to invent a new institution to cope with it." He mentioned specifically

> . . . our responsibility to control the effects of our own scientific experimentation. For as science investigates the natural environment, it also modifies it--and the modification may have incalculable consequences, for evil as well as good. . . . The Government has the clear responsibility to weigh the importance of large-scale experiments to the advance of knowledge to national security against the possibility of adverse and destructive effects. . . . As we begin to master the destructive potentialities of modern science, we move toward a new era, in which science can fulfill its creative promise and help bring into existence the happiest society the world has ever known.

The responsibility to assess technology is recognized, but assessment so far has been a haphazard affair, initiated by the Rachel Carsons and Ralph Naders or by episodic events such as oil slicks on beaches. These stimuli for assessment have been embarrassing to both the technical and political communities because the lack of control over applied science is dramatically demonstrated.

In this climate, the U. S. Congress found that its conventional sources of information were inadequate to the task of adding scientific and technical ingredients to the legislative decision-making process. Essentially, the function of the Congress is to integrate all the elements of national life, with their varying and often conflicting desires and the complex of social, legal, economic, political, and institutional values. The result is the formulation of national policy, and the setting of priorities and timetables via the three major activities of authorization, appropriation, and overview. The success of the integration and compromise is judged directly by the people at the polls.

The Congress is not bound to programs during their execution and thus is in a position to critically evaluate the results, in contrast to the legislative intent. The fact that all Representatives and about one-third of the Senators stand for election every two years presents an opportunity for relatively rapid response to public opinion, compared to an administrative program that may go on for four or eight years and become entrenched in bureaucracy.

Information about science and technology is not present in the Congress in the most usual input form, i.e., the personal experience and training of Representatives and Senators. Few scientists and engineers have been elected from an active professional career, and there is no sign the number will increase. Thus, an information transfer process is more important in technical matters than in other areas of knowledge.

The Congress certainly does not lack for information. The openness of the legislative process provides a great variety and number of channels for facts and opinion. The public hearing is common to almost all legislative considerations.

The Federal agencies advocate and defend their programs in hearings, briefings, and continual liaison with the members and committees directly concerned. The Office of Science and Technology in the Executive Office of the President was established by reorganization plan specifically to coordinate executive branch R&D and to report to the Congress on science policy matters.

Special commissions are sometimes created by the Congress to examine a policy problem in depth. For example, the National Water Commission, the National Commission on Technology, Automation and Economic Progress, and the Commission on Marine Science, Engineering and Resources are Congressionally constituted to report both to the President and to the Congress.

Industries and trade organizations often present testimony at hearings and carry on extensive information and lobbying activities when legislation or government programs particularly affect their interests. Constituents and local businesses make specific inquiries to their representatives and carry opinions on R&D matters that have obvious political impact.

Universities and nonprofit institutions provide data and interpretive studies in the normal course of their activities.

Professional technical societies can be a most valuable source of advice and information for the Congress. In contrast to the other sources, the broad membership allows professional, rather than employment-related views, to come to the surface. Their tax-exempt status need not be jeopardized, nor are they required to register as lobbyists.[1] The simple conditions are that Congressional liaison not be a major portion of their activities and that the Congress has invited their testimony.

The technical societies need not be expected to present a consensus view on issues; that might engender internal strife. Rather, they may provide a forum for discussion of science policy issues just as they have historically for the presentation of research results. This new opportunity for professional societies is well worth adding to the traditional functions of publication, national meetings, and education.

As the need for science information has become recognized, the legislators have erected a variety of transfer mechanisms and channels. Some members have personal science advisory committees. The political parties have organized scientists and engineers to assist them.

In the past few years, professional staffs of committees have been reinforced with persons of mature technical experience. Hearings and briefings are often purposefully arranged as adversary proceedings to present a variety of viewpoints. Ad hoc advisory panels are assembled to deal with a particular subject or to provide general advice.

In 1963 and 1964, a number of studies and proposals were made relating to science advice for the Congress ranging from Nobel laureates in residence to a "think tank" organization. The Subcommittee on Science Research and Development (chaired by Mr. Daddario) of the House Committee on Science and Astronautics held hearings and issued a report, "Scientific-Technical Advice for Congress: Needs and Resources."[2] Bills to create a Congressional Office of Science and Technology were introduced in both Houses in the 88th Congress (S. 2036 and H.R. 8066). Another bill (H.R. 6866, 88th Congress, 1st session) sought to establish a small science advisory staff in each house.[3] The House Select Committee on Government Research conducted extensive studies on the methods by which Congress could improve its handling of Federal science affairs.[4] Senator Clinton P. Anderson reviewed the difficulties of setting up a body of highly trained technicians solely responsible to Congress in an article in *Science* magazine.[5]

The conclusion of these deliberations was that a bridge between the Congress and the scientific community was preferable to an "in-house" cadre of practicing scientists and engineers. The breadth of legislative problems and the rapid change of specific interests militated against retaining subject matter specialists as advisers. Instead, the need was correctly deduced for a new function to excise the scientific questions from legislative issues, to search the literature and minds of the scientific community for relevant answers, and to interpret and distill the pertinent information for easy use by the legislators.

The means of establishing the function was found close at hand. The Legislative Reference Service (LRS) is a fifty-three-year-old organization with the Library of Congress. It is divided into ten divisions that provide research and information on all fields of knowledge related to legislation. The staff numbers over 300 persons, ranging from senior scholars to clerks who retrieve factual data. The specialists are thoroughly qualified and recognized in their academic fields, and, more importantly, they are skilled at interpreting and communicating technical information to the Congress.

LRS works only for the members and their committees and staffs. Projects are performed on a confidential basis, and usually the resulting contributions in printed form do not indicate their origin in the Library. Partisan political influences are carefully avoided under the direction of the Joint Committee on the Library.

Into this existing organizational framework, the Congress introduced eight new positions by appropriation action for fiscal year 1965 "to meet the increasing requests for service in the scientific and technological specialities." These positions were augmented with existing funds and organized as the Science Policy Research Division (SPRD) in October of 1964.

At present, SPRD comprises ten senior professionals, ten junior researchers and assistants, and five clerk-typists. The senior researchers have backgrounds in science and engineering, which gives them a legitimate standing in the scientific community. Obviously not all disciplines can be covered by so small a staff, but choices are made to correspond to the principal issues at hand, i.e., space, transportation, environmental pollution, oceanography, etc.

In a typical year assignments will be handled for fifty different committees and subcommittees, although the majority come from the House and Senate space committees and a few other technically oriented legislative panels. About 400 (of the 535) members will be served as individuals.

To be effective as a complete service organization, a reference function is also performed: several thousand inquiries are answered every year. This factual data retrieval has another purpose, however, in that it acquaints members with the division and often leads to policy studies. A research project may be initiated by the SPRD in anticipation of an issue likely to come before the Congress.

The objective of the Science Policy Research Division is an increased understanding by the members of Congress of the workings of science and engineering. Scholarship, or completeness of research, and objectivity, or the absence of advocacy, are the criteria of performance.

SPRD is a bridge to a diversity and plurality of information and opinion. It is not a primary source of knowledge. Nevertheless, the way in which the gathered facts are presented to the legislator often does impart a summary judgment, a responsibility which is carefully recognized and accepted.

Our experience to date is that Congress does not need advice and that judgments on technical matters must continue to be rendered within the conventional legislative process. It is not necessary or desirable for the Members to become technical experts. What many of them have become is sensitive to the capabilities and limitations of science and technology. As the National Academy of Sciences report "Applied Science and Technological Progress" states:

> Congress should not attempt to second guess the experts on technical appraisals, but it does have the responsibility to convince itself that the experts have asked themselves the right questions. . . . In appraising the situation, it is important for Congress to listen to the skeptics as well as the enthusiasts, and to ask the enthusiasts to answer the arguments of the skeptics. Laymen can learn a great deal from the confrontation of experts even when they do not understand the details. Especially in applied science and technology, priorities and goals can be established only through a multidimensional interaction between scientists, technologists, public servants, and the general public.[6]

Thus, the critical need of the Congress is to acquire the capability for assuring that competent and timely assessments are done and for transferring assessment results into a form applicable to legislative decisions. Our technological problems are part of our political problems, with social, personal, and economic costs and benefits.

The Congress is the political assessment body in our society and must have the output of technology assessments in order to do its job.

In operation, the Congressional technology assessment capability will initiate assessments, search out ongoing studies, structure hearings to bring out adversary views, monitor assessment functions in the executive agencies and the private sector, and assess the assessors. A relatively small group of professionals (generalists from a number of disciplines) plus a modest appropriation for contract studies will suffice. This capability may also carry out assessments when appropriate, but a bridge to major efforts elsewhere is the primary intent. Since it is expected that technology assessment will become an integral part of scientific and engineering activity, the Congressional function will be a logical extension of the present work of the Science Policy Research Division.

The fact that applied science permeates the entire fabric of society argues against a single assessment institution and argues for the development of assessment mechanisms in a wide variety of government agencies, universities, nonprofit organizations, and commercial firms.

The location of government technology assessment solely in the Executive Branch would appear to generate impossible conflicts with the necessary promotion and championing of programs and the aggressive administration of the law. The Executive Branch determines for its own purposes whether something is worth doing, but the Congress determines whether this is good for the whole country. The establishment of a fourth (evaluative) branch of government, suggested by Golovin,[7] may present an interesting alternative when fully developed. One objection, however, is that such a centralization would interfere with the necessity (in my mind) that assessment be a part of the innovative process wherever it goes on.

The case for a Congressional assessment capability is supported by the proceedings of a seminar held in September, 1967.[8] Some of the conclusions from that discussion are the following:

1. Questions of concern to the Congress must be assessed to meet the legislative schedule, i.e., before decisions are made.

2. The Congress must be in a position to continuously challenge the Executive Branch as to the consequences of its programs.

3. Assessment must proceed in an open forum stimulating wide

public discussion. The Congress can provide such an environment.

4. The widest possible base of information and opinion must be accessible to assessment projects. The Congress could command this knowledge.

5. The Congress would be more attentive to assessment results if they were presented via a Congressionally chartered organization.

6. The Congress must be convinced that the experts have asked one another the right questions.

7. The political decisions affecting the future of technology rest with the Congress.

8. The Congress is sensitive and rapidly responsive to the people and is immediately accountable to the electorate.

9. The feeling that applied science is under control (through Congressionally monitored assessments) will restore public confidence necessary to a risk-taking progressive society.

10. The needs of the Congress for assessments results would assure that the necessary funds for these activities would be provided.

NOTES

1. U.S. House, Committee on Science and Astronautics, Scientific-Technical Advice for Congress - Needs and Sources, 88th Cong., 2nd sess., (Aug. 10, 1964), p. 63.

2. Ibid,; also Hearings before the U.S. House, Committee on Science and Astronautics, Subcommittee on Science and Research and Development, Government and Science, 88th Cong., 1st sess. (Oct. 15-Nov. 20, 1963).

3. U.S. House, Committee on House Administration, Subcommittee on Accounts, Hearings on HR 6866 and HR 8066, 88th Cong., 1st sess. (Dec. 4, 1963).

4. U.S. House, Study Number X--Part II, Staff Resume of the Activities of the Select Committee on Government Research (Feb. 18, 1965).

5. Clinton P. Anderson, "Science Advice for Congress," Science, CXLIV (April 3, 1964), p. 29-32.

6. U.S. House, Committee on Science and Astronautics. Report Applied Science and Technological Progress, by National Academy of Sciences (1967), p. II.

7. U.S. Department of Commerce, The "Evaluative Function" in Government, by Nicholas E. Golovin, National Technical Information Service Publication, PB 180 359 (1968).

8. U.S. House, Committee on Science and Astronautics, Subcommittee on Science, Research, and Development, Technology Assessment Seminar, 90th Cong., 1st sess. (September 21 and 22, 1967).

DISCUSSION OF TECHNOLOGY ASSESSMENT AND THE CONGRESS

Mr. Carpenter provided an extremely accurate introduction to the discussion when he noted that "this is a pretty provocative subject." In asking those present to indicate their opinions concerning the type of information Congress needs, he extended upon the remark in his paper that Congress does not need advice by commenting,

> What I should have added is that Congress obviously does not need or will not take advice on how to vote, even on fairly specific technical matters such as the accelerator at Weston or Mohole. Congress won't abdicate the final judgment in these matters and shouldn't. No one of these decisions is purely technical: you can't separate the social, political, and economic aspects from the technological problem.

What is needed is reliable, complete, and objective information. Perhaps the study of the Congressional role in technology assessment will lead to a search for new institutions that will enable us to "preserve a democratic pattern of decision-making and yet take full and optimum advantage of what is indeed a very technological world." Mr. Carpenter viewed the problem as that of providing a "bridge to Congress" which will make the greatest possible use of technological knowledge and ongoing assessments in other sectors of the society.

LEGISLATIVE REFERENCE SERVICE

The discussion of the Science Policy Research Division of the Legislative Reference Service (now the Congressional Research Service) of the Library of Congress, which was included in the prepared paper, was expanded by some of the participants. A Congressional Aide wanted to make clear the fact that the LRS does not take an advocacy position but rather, in an impartial manner, lists the questions of interest in any particular matter. Its study then provides the foundations for Congressional committee hearings. In view of comments made at first seminar (See Chapter 2.) concerning the role of experts in the assessment process, a university researcher wanted to know how the Science Policy Research Division of the LRS, which is made up of scientific specialists, avoids the "specialized expert" problem. It was pointed out that the LRS includes specialists in a wide range of fields and that the Science Policy Research Division has "many opportunities to use public administration people, political scientists, social scientists, and economists within LRS." A Congressional Aide mentioned that the LRS has the authority to hire outside experts for specific purposes. The exercise of this authority leads, in his view, to a flexibility in the operation of the LRS, which makes its studies particularly valuable. Further, it was reemphasized that it was not the purpose of the LRS to make judgments; the hallmarks of the LRS, according to one of its staff, are "scholarship and objectivity." He explained:

> We try to get all the relevant opinion pulled out of the literature and pulled out of the minds of persons around the country. . . . We try to present it [the information obtained], interpret it, analyze it, and reduce it to a useful form in as objective a manner as possible and we use internal review and some external review to assure that our own prejudices and advocacies haven't slipped into the final product. Obviously, we do this successfully sometimes; not so, other times.

CONGRESSIONAL COMMITTEES AND HEARINGS

The discussion turned to a consideration of Congressional hearings and almost immediately became highly animated. One participant, using the Joint Committee on Atomic Energy as an example, charged that some Congressional committees deliberately prevent certain views from being heard. The difficulty, he contended, arises when the committee "predetermines the conclusion of the hearing; then it sets up the structure of witnesses that will justify the conclusion it predetermines." A professor agreed and added that the problem of obtaining a hearing is by no means limited to one or a few individuals but is, in fact, quite widespread. Even a committee staff member agreed: "You are quite correct when you say that many times a committee will decide what they want to find out first and then flush out the points of view they want to use." An agency official noted that the adversary process used in committee hearings provides an opportunity to "stack the deck; it does seem to be very difficult to ensure that in every case the adversary process will be an open one." A Congressional staff member took issue with these points of view; he believed that committees are fair and that they adopt a policy that anyone can be heard. This, he said, is made clear in the public announcement of hearings. Another Congressional Aide noted that when all witnesses are not permitted to testify, the cause is often the limited time of Senators rather than bias. The strong differences of opinion among those present over the partiality or impartiality of hearings are perhaps best exemplified by this brief exchange:

Congressional Staff Member:

> We're fair. We don't shut people off.

Participant:

> Some of you must deal with a different Congress than I do.

The controversy continued when a speaker remarked that he regularly finds in the announcement of hearings a requirement for submission of a statement in advance, before the committee decides who is to testify. He concluded that "we need, in whatever the structure, an opportunity for both sides, or all sides, of the issue to be included, with the requirement that it all be made public." But a Congressional Aide answered that the requirement of prior submission of testimony exists because some statements are clearly "way out" and the product of quacks. He continued: "We can't just let somebody come in with a statement. We have to, in some way, get some sort of perspective before we put on our witnesses." Another committee staff member pointed out that experience with the Joint Committee on Atomic Energy (JCAE), which had provided most of the examples up to this point in the proceedings, cannot really be generalized. The JCAE is the only joint committee with legislative authority and as such is a special case. In general, he held, the bicameral system allows interested parties two opportunities to obtain a hearing and thus provides some protection against exclusion from the proceedings.

An agency official pointed to the proposed antiballistic missile (ABM) system as an issue outside the jurisdiction of the JCAE, on which it was difficult for scientists opposed to the system to gain a hearing. The scientists, he said, feel that "they were excluded by the Committee Counsel and that they were systematically disallowed from testifying." Another agency official objected to this example; he claimed that controversies over weapons systems cannot be considered as purely scientific questions. "When you talk about weapons systems, you're talking a different game," he said, to which a speaker replied, "I think we've got to stop that game. We're doing things more and more in the name of national defense."

A professor claimed that "certain Congressional committees, and I would define these as being mission-oriented committees which have mission-oriented agencies as their clients, are every bit as "gung-ho" for technology as the executive sponsors of the program." It is thus vitally important that all members of Congress (and the public) as well as those members of interested committees, receive information from such sources as the Legislative Reference Service. A speaker pointed out that it is not true that all committees demonstrate this protechnological bias and cited the example of Representative Fountain's Committee and its relations with the National Institutes of Health. As for providing information to all Congressmen, an LRS staff member noted that the reports of the LRS are available to any Congressman who desires them, and that distribution is not limited to the committees which request particular studies.

ADEQUACY OF ADVICE

Interest then turned to the question of the type and adequacy of advice received by Congress. This topic had been discussed in Chapter 2 with no real conclusion or consensus having been reached, and the conversation continued in a similar vein. The lack of agreement should probably not be taken as an indictment, but rather as an indication of the difficulty inherent in the problem.

A professor posed the question: "Will independent scientific study groups in different agencies using the same methodology and science come to the same conclusion [on any given issue]?" He went on to answer that he, for one, believed that they will. Another participant agreed, but a third, a staff member of the LRS, challenged the view. He felt that the nature of the body performing a study played a large part in determining what information is used in the analysis (and thus in determining the conclusions). He presented the example of a Department of Agriculture-Forest Service study of fire prevention methods in which "a great deal of research was buried simply because it negated the whole function of the Forest Service." Thus, some bodies of fact are hard to find, "even for the LRS," he added.

There was some comment about the role of professional societies in technology assessment. One speaker contended that societies should take a more active part in the assessment process and another, an agency official, agreed, although he noted that "you are asking for a great deal in asking the professional societies to step outside their traditional role of being oriented just toward the furtherance of their scientific disciplines." A Congressional staff member thought that it was important to make a "distinction between those subjects which are within the competence of the people in the society and broad social questions." He believed that the American Physical Society (APS), for example, ought not make policy statements on problems such as "poverty in the South." Decisions about the needs of society are in the province of Congress alone and not in that of the APS or anyone else. This topic has provoked much controversy within scientific societies, but for the purposes of this seminar session, the participants agreed that this brief discussion was sufficient.

An agency official thought that before anyone can really understand the type of "bridge to Congress" that would be most successful, it is necessary to "disaggregate the term 'technology assessment.' " He identified two kinds of assessments. The first involves an appraisal of effectiveness in achieving specific program objectives, and the second involves "looking at ecological consequences" of technological applications. He classified the two assessment processes as the examinations of "sought consequences and unsought consequences." For example, if one were concerned with removing stains from clothing, one kind of assessment would consider whether a given product will in fact effectively remove the stains. The other would examine the overall consequences of using the product (water pollution, for instance). The two types of assessments are totally different, and it is quite likely that different assessment mechanisms are appropriate for each. A professor, however, claimed that "this is the situation we have now and that's why we're in such trouble." It is precisely because assessment is disjointed that problems exist. Thus, stain removers are, in fact, evaluated on the basis of how well they remove stains independent of the environmental effects which their use can produce. Another agency representative agreed with the first speaker; he reemphasized the point that feasibility is a question of a different order than the "trade-offs" involved in comparing benefits and risks and that "if you don't make these distinctions and try to answer these questions in one fuzzy blur, you'll find that you can't answer them." All participants felt that it was crucial to examine overall consequences of the use of technology; the point of contention which was left unresolved, was whether conceptually disaggregating the assessment process helps in achieving a complete analysis or causes people to ignore important parts of the problem.

A NEW GENERATION OF ENGINEERS

Because at least some of the information which Congress needs to make decisions concerning technological applications must come from those who are technically trained and competent, a professor suggested that in the long run "what is needed is the creation of a new generation of engineers who are trained so that in the forward design of the systems that they are creating they build in certain retrievable values, so that when the system accomplishes its primary mission it has either a receptive value or can find some recoverable value." He charged that engineers are trained irresponsibly. The very people who will eventually become contract administrators, who will help to decide which systems should be adopted to obtain certain ends, are not currently taught to be aware of social values. It was noted that a prototype of the training system for the "new engineers" has been developed and is currently operated at Stanford University. An agency official suggested that the "new generation of engineers" might be called "public policy engineers." However, not everyone was enthusuastic about the concept of new systems for training engineers. The following exchange demonstrates the reaction of one participant to the creation of "new engineers":

Congressional Staff Member:

> They won't be engineers anymore if you do that.

University Professor:

> Change the word--it's semantics.

Congressional Staff Member:

> That's not true. If you take this thing too far, you end up with people who are not going to do the nuts and bolts of engineering anymore.

The current engineering profession was discussed by two other participants. A professor, noting that the engineering profession is the vehicle through which scientific advance leads to technological change, wanted a clarification of the meaning of engineering:

> I must confess that despite somewhat diligent inquiry into this, I've never been able to quite figure out what the concept of engineering embraces. It is only concerned with affixing the nuts to the bolts, or is it, on the other hand, a profession? If it is a profession, are there any ethical considerations which are relevant to the manner in which engineers behave as they translate science into technology or are they, to put it bluntly, prostitutes?

No direct answer was forthcoming.

An industry representative contended that the current use of the word "engineer" in the sense of one who operates an engine was unfortunate. He said that the European usage, in which the term means one who exercises his ingenuity for private or public good or destruction, gives a better understanding of what an engineer really is.

THE PUBLIC INTEREST

The most important aspect of technology assessment, according to one agency official, is the attempt to take an "ecological viewpoint" or a "general systems viewpoint." "What human society is about is the public interest and not the interests of science or the interests of technology." This immediately provoked the comment that "there is a difficulty in defining what the public interest is." Several participants volunteered to try to devise a definition. One, an agency official, thought that Congress decides the public interest through its actions. Another speaker defined the public interest as "opposition to any increment toward destroying the human race." Still another felt that questions about whether decisions are made in the public interest are "silly. . . . [in] a democracy the most that you have to do is make the facts available to the public." If the public chooses not to take effective action based on the facts, then the government can do nothing. That is, the government should not provide protection to the public if the public does not want it. (This point is further elaborated in Chapter 4.)

Need for Supplementary Institutions

To one participant, the primary question to be addressed is, "How do you organize social criticism effectively so as to affect public policy?" Organized social criticism can be quite potent in producing action. President Johnson, for example, was "brought down by the operation of intense. . .social criticism that affected institutions all through the society." With respect to technology assessment, one must determine how social criticism may be organized in the appraisal process and what type of organization will bring out the development of beneficial technologies without taking risks but also without destroying the confidence of entrepreneurs. The speaker pointed out that:

> The problem becomes one of looking at your institutions. We can talk all night about improving what we've got, like a new generation of engineers, which I don't have any fight with, except that I won't be around by the time it is produced, and in the meantime God knows what will happen. It seems to me that the effective way to proceed is by developing supplementary institutions with the idea in mind that they will not just be made up of engineers or scientists, but will be multidisciplinary.

He proposed an independent institute of technology assessment. The independence of such an institute would allow it to avoid one of the problems he sees in the operation of the Legislative Reference Service:

> What I suspect is part of the difficulty of the LRS is that that. . . much of its traffic consists of being responsive to questions on which answers are wanted tomorrow. To remove the mechanism from the consequences of erratic and random demand would be a constructive thing.

An independent institute could provide a public, as well as a legislative, service. Financed by Congressional charter or by an endowment, the institution would serve as an independent voice of "exposure, disclosure, criticism, prediction, and early warning."

When asked by a Congressional Aide "Who would listen?," the institute's proponent admitted that he could not answer that question and that the question was one which "troubles me very much, because I think it says a great deal about the quality of intelligence in the society, and maybe we deserve it." A Congressional staff member asked why the National Academy of Sciences and the National Academy of Engineering could not serve the purpose of the proposed institute. The answer was that these are membership organizations and, as such, have a corporate role that is not necessarily consistent with the goals of a technology assessment institute. An agency representative thought that the suggested institute already exists in the universities; but it was pointed out by a Congressional Aide that university efforts are fragmented. It was further noted that, at least at present, the university is not publicly recognized as the institution which fits this particular problem. One professor commented that the proposed assessment organization was "the institutionalization of the Rachel Carsons and the Ralph Naders," but he was assured that the intent of the technology assessment institute was not the "construction of an asylum."

Most of these present agreed that the proposed institute would be a positive step toward the improvement of the technology assessment process, but there were some questions concerning the mechanism whereby such an institute could report its findings. One participant, for instance, was concerned with "how this information-generating capability, from wherever it comes, get introduced into the political decision-making process." But, he continued: "It seems to me that one usefulness of [such] an institution would be that it might well create a national constituency against the kind of politics that we are accustomed to, which could help in the decision-making process. The problem of the "bridge to Congress" still exists, in his view, and he concluded:

> Our decisions are made through representative government, and it's likely to continue that we will be making decisions through representative government. Therefore, the linkage which Dick [Carpenter] proposes is still an important one.

CHAPTER

4

THE ADVERSARY PROCESS IN TECHNOLOGY ASSESSMENT

Harold P. Green

My purpose in this paper is to show the importance of introducing an adversary process into the technology assessment mechanism in order to assure that the assessment function will properly and effectively protect the public interest.

At the outset, let me define my terms. When I speak of an adversary process I am not suggesting a formal adjudicatory process in which opposing parties contend through legal mouthpieces with a decision made of the record. Rather, I suggest only that a mechanism be developed that will permit and facilitate the articulation in public of all relevant facts, pro and con. And when I speak of technology assessment, I do not encompass the assessment of the full potential range of the social consequences of a technology. Rather, in this paper I am concerned with technology assessment solely from the standpoint of identifying and controlling those attributes of a technology that adversely affect basic individual rights that have traditionally been protected by the legal system, i.e., specifically those incidents of a technology that may threaten the health, safety, and security of the public. I believe, incidentally, that the aspect of technology on which I am focusing is its principal element and the very *raison d'etre* for the present public discussion of technology assessment.

---

*Presented March 19, 1969.

Most of the discussion of technology assessment taken place to date has assumed the need for a new governmental assessment institution that would identify and quantify the benefits, costs, and risks of technology, then strike a balance among these factors, and pass on its conclusions and recommendations in predigested form to Congress, as a predicate for legislative action. This approach is reflected in Congressman Daddario's bill and was followed in Mr. Carpenter's paper. Mr. Carpenter's view was that Congress, as the political assessment body in our society, must have the output of technology assessment bodies in order to do its job properly. I have no serious objection to this idea. Subject to my concern that Congress already is the recipient of more information than it needs and can handle effectively, even with the aid of computers, I have no doubt that Congress will inevitably benefit from receipt of the balanced, objective views of any responsible body. But I do believe that this approach is not adequate, would not be effective, and is in derogation of the appropriate role of Congress and the public in policy formulation.

The assessment function is properly concerned with balancing the benefits of a technology to the public against its costs (including risks) to the public and emerging with a conclusion as to what government's role with respect to that technology should be. Unfortunately, the benefits and costs (including risks) do not fall upon all segments and members of the public proportionately. There is, I suppose, a high degree of correlation if we are talking about lawn mower technology; but if we are talking about detergents, pesticides, or the supersonic transport plane, it is clear that those who enjoy the benefits of these technologies will not necessarily be the same persons, or in appropriate degree, as those on whom the costs (including risks) will fall.

Since the issue is one of benefits to the public versus costs (including risks) to the public, the focus of technology assessment should be to arrive at a conclusion as to what costs (including risks) the public is prepared to assume in exchange for what benefits. In our democracy, such decisions cannot appropriately be made by an elite body of specialists and generalists (who are specialists in technology assessment). They should be made by the public itself expressing its views through its elected representatives in the Congress, who are accountable to their constituents. This requires that the entire assessment process take place in the open with full articulation, in language the public can understand, of the benefits and costs (including risks). In short, I do not agree with Mr. Carpenter that there is any viable distinction between political assessment and technology assessment. Technology assessment is not an appropriate function for experts; rather it is a process which should be performed entirely at the political level. Those who question whether the public and the Congress have the competence to make the necessary sound assessments express a lack of faith in the democratic process to cope with modern and future technology, and if we act on the basis of such lack of faith we have a different ball game. In my view, the basic problem is to compel scientists and technologists to present the issues to the public in the language of ordinary public discourse rather than in the esoteric jargon of their disciplines. If this is done I have no doubt of the efficacy of the democratic process.[1]

Most public discussion upon technology assessment to this date has ignored a fundamental point. There is never any lack of articulation of the benefits of a technology. Every technology has powerful vested interests--private and frequently governmental and political--which can be relied upon to press the benefits to the technology assessors. The problem is that the negative factors and the risks are never fully or even adequately articulated. The inability of the negative factors to surface effectively is attributable to a variety of circumstances. In some cases the risks are totally unappreciated until a later date; in other cases, there may be an appreciation of possible risks that have not yet been demonstrated to be real. The proponents of the technology may always be counted upon to minimize or suppress the risks. Although the proponents are usually well-organized and well-financed, those who seek to advance the negative factors tend to be rather disorganized and to lack resources. Not infrequently--particularly in the case of government-sponsored technologies--it is difficult for the opponents to obtain relevant and adequate information about risks, and even more difficult--because the experts who are privy to the relevant information are usually pro-technology--to obtain experts to assist them in formulation of their contentions. The natural consequence is that the opposition is forced to state its case using information which is incomplete or not wholly accurate, and, therefore, is easily discredited. Frequently the establishment seeks to discredit the opposition ad hominem. This exacerbates the situation by forcing the opposition to take an extreme position, which makes it even easier for the proponents to discredit their contentions.[2]

The basic problem of building an assessment institution is, therefore, providing a means whereby the negative factors, particularly the risks, will be vigorously, effectively, and responsibly pressed upon the decision-makers in a manner that will permit the Congress to make its own judgments and that will permit the public to make its own judgments, so that its views will become known to the Congress.

In a nutshell, the problem is to give equal time, opportunity, and attention to the negative factors.

It is in this respect that I think Congressman Daddario's initial proposal for creation of a Technology Assessment Board is deficient. The Board's findings, conclusions, and recommendations ex cathedra will not reflect what the public wants and what the public fears. The very expertness and authority of the Board will detract from public discussion and debate. I have the same concern about Mr. Carpenter's proposal. In both cases, incidentally, the suggested technology outputs will not be considered by the Congress in isolation. Congress and the public will still be faced with a barrage of propaganda from the technology's proponents, and we may be certain that they will attack and seek to minimize the negative considerations brought to light in the course of these assessment processes.

These considerations lead me to the conclusion that what is needed for the technology assessment function is an agency which would act as a responsible devil's advocate or technological ombudsman and play the role of adversary in the Congressional and public forums. This should be an agency charged solely with the function and responsibility to probe for the negative factors, to identify these factors, and to press them vigorously upon the Congress and the public. The agency should be either an agency totally independent of the government or one which, like the General Accounting Office, functions as a part of the Congress. Between these alternatives, my preference is for the latter, since the agency should have the right to have full access to all relevant information within the control of the government.[3]

In discussing the manner in which such an agency would operate, it is necessary to distinguish between two classes of technology. On the one hand, there is technology that is essentially private and that is developed and introduced primarily as a consequence of private, profit-seeking investment. With respect to such technology, government's role is typically passive until problems arise that require the government to take action in the public interest. On the other hand, there is another class of technology that is developed primarily as a consequence of government investment. In this case, government has a strong affirmative interest in development of the technology and also in its subsequent practical application.

In the first of these cases, the marketplace operates as a continuous technology assessment mechanism. A technology will not be developed or introduced unless its sponsor senses that there will be a profitable return on investment. The price that buyers are willing to pay reflects their assessment of the hazards that may be incident to the technology's products.[4] Both the sponsors of the technology and its consumer-users also assess the potential costs to them that may result from liability to others arising out of such hazards. Liability itself results from an assessment process by the courts since the social utility of the activity producing the injury to the plaintiff is frequently weighed (sometimes explicitly) against the deterrent effects of imposing liability.[5] The estimated costs of potential liability may be a deterrent to one who considers developing and introducing a technology or using a technology or its products. Liability insurance, which pools like risks and translates uncertain potential liability into fixed periodic premiums, serves to offset this deterrent, but if adequate insurance is not available or is available only at prohibitively expensive rates, the deterrent remains.[6] These private assessment mechanisms involve a mix of incentives and deterrents that operate as people pursue their own self-interest. These mechanisms flash a green light or a red light that serves to control the rate of development and introduction of technology. It is only when the green light has been flashed as a signal for introduction of a technology that the problem of technology assessment passes into the hands of governmental institutions. Even after the technology has been introduced, government assessment does not occur until problems arise which seem to warrant governmental action.

The initial step in governmental technology assessment occurs when the legislature considers these problems to determine whether social control is required to protect the public against the technology's hazards and, if so, the form such controls should take. The mere existence of obvious problems does not mean that the legislature will act. The legislative process involves considerable inertia and is at best an uncertain and lengthy affair. The proponents of social controls battle the vested interests who stubbornly resist control.[7] Typically, the legislative struggle involves two issues: 1. Are there hazards which justify social control? and 2. Will the form of social control stifle the technology and deprive the public of its benefits? Where legislation results, it represents the striking of what the legislature believes is an appropriate balance between benefits and risks.

In many cases the social controls are implemented under statutory standards by an administrative agency. The agency's sole function in this respect is to protect the public interest under the statutory standards. These standards usually reflect the legislature's conclusion that the technology is useful and beneficial and should be controlled in such a manner as to preserve its benefits. As the administrative agency functions, it is required to assess the impact of its proposed regulatory actions in terms of benefit and risk. Because its institutional bias is in the direction of protecting the public interest, the regulatory agency generally functions as an adversary of the industry it regulates; i.e., its normal functioning imposes social controls, at the expense of increasing costs and reducing benefits, in order to maximize protection of the public against the industry's hazards.

We have, then, in this case an existing structure for governmental technology assessment. Obviously, it does not work perfectly. Legislative judgments are often inappropriately biased or otherwise unsound. Administrative agencies make mistakes and, as students of the administrative process have long recognized, they tend to develop unwholesome affinities to the industries they regulate and may become, in effect, their captives. Such deficiencies are inherent in the political and administrative process and are subject to corrective action if the public is aroused. On the other hand, a more difficult problem is raised by the fact that the assessment institutions frequently lack sufficient information and knowledge to function with optimum effectiveness. These institutions and the public frequently do not become aware of hazardous conditions associated with a technology at a sufficiently early date, until after considerable harm has occurred and strongly entrenched vested interests have come into being. What is needed is an early warning system to trigger public and political awareness of potential hazards at a much earlier time.

The type of devil's advocate agency I have suggested is ideally suited to overcome these deficiencies. Its functions would include that of identifying possible hazards and ascertaining the extent to which they are real hazards. The agency would identify areas in which regulatory bodies should prudently force the technology to slow down and areas in which further research may be necessary. Hazards and potential hazards, when these are identified, would be vigorously publicized and pressed upon the Congress and the cognizant government agencies. Congressional inertia would be subject to the pressures of public opinion; foot-dragging by the administrative agencies would become more visible and subject to correction. A more substantial burden of proof would be thrust upon the sponsors of the technology. I do not suggest that this approach will result in a perfect technology assessment function. In a democracy, one cannot seek or expect a perfect solution to every problem; one can ask and expect only that the facts will be made known so that political forces can operate to produce the solution the public demands.

The second case,[8] that of government-sponsored technology, is quite different. These technologies develop with government investment, which is in no way related to the forces of the marketplace; indeed they develop in defiance of the market since government investment is made for the very reason that the market does not provide incentives for development of the technology on the time scale government believes is necessary. Thus the "deep pocket" of the government supports technology development merely because desirable benefits are foreseen, even though there are no market incentives, while none of the restraints and deterrents, which are present with respect to privately developed technologies are operative in this case.

It is apparent, moreover, that as the technology is assessed at various points (usually in connection with authorization of the program and appropriation of funds[9]) as it passes from the hands of the specialists-sponsors within the agency through the less specialized offices in the higher echelons of the Executive Branch to the generalists in the Congress, there is a natural tendency to minimize the existence and significance of any potential hazards associated with the technology. I do not mean to suggest that hazards are callously ignored. On the contrary, they are usually carefullv explored and considered by the sponsors of the technological development. But the existence of even substantial potential hazards rarely chills the enthusiasm of the mission-oriented agencies eager to demonstrate that what is possible is feasible and will bring immense benefits to the public. Hazards and risks are characteristically rationalized away. The rationalizations take one or more of the following forms:

1. The hazards are not as serious as might appear; at least there has been no demonstration that they will indeed be harmful.

2. Even if there may be some hazard or inconvenience to the public, this may be "tolerable" or "acceptable" in view of the enormous benefits the public will receive from use of the technology.

3. We are conducting research to learn more about the potential hazards and research and development to provide a technological fix to eliminate or reduce the hazards.

4. In any event, there is no need to worry about possible adverse consequences of the technology until its feasibility has been established.

5. Obviously, the government will permit use of the technology only subject to appropriate controls to assure that the public will not be injured, and if such controls are not adequate to protect the public, government will not permit use of the technology. QED.

One need only consider the history of government development of such technologies as atomic energy,[10] weather modification, the supersonic transport plane, and various military technologies such as biological and chemical agents to appreciate that existence of very substantial hazards has not deterred development of such technologies because of precisely such rationalizations. It is clear, moreover, that such rationalizations have been accepted at the higher levels of technology assessment, i.e., the Executive Offices, the cognizant Congressional committees, and the Congress. The impulse to convert science into technology that will benefit to society seems to be irresistible at every level. In part, at least, this may be attributable to the fact that information about the adverse consequences reaches the higher and more generalized areas of technology assessment, if at all, in a highly distilled form that does not facilitate discussion and debate.

Although we do not have too much experience, outside of the military and space areas, with the process through which government-developed technology is introduced into practical application, some generalizations are possible. It seems inherent in the American governmental and political process that government will have an irresistible impulse to see technologies developed by it put to practical use. For one thing, bureaucrats and politicans cannot lightly contemplate the prospect of having spent tens or hundreds of millions of dollars on a project which turns out to be a white elephant in the form of a feasible but unusable technology. Can it be doubted that weather modification will be practiced when the technology is developed even though interests of a substantial number of people will be substantially and adversely affected? Or that the supersonic transport plane, when and if finally developed, will eventually be permitted to fly in a manner which will result in effects which many members of the public regard as intolerable? Again, I do not want to be read as suggesting that government callously will foist hazardous technologies upon the public. There is really no such thing as a clear-cut dichotomy between "safe" and "hazardous." This is an area in which there is no black and white. Whether or not something is appropriately safe or unduly hazardous is an issue which lies in a gray area and the resolution of which is essentially a matter of judgment on which reasonable men can differ depending upon their outlook and biases. All that I suggest is that those who have a vested interest in a technology will inevitably be more relaxed about hazards than those who don't give a damn about the technology but who are primarily concerned with the public safety.

We do have substantial experience that is instructive in one area. The Government has for many years sponsored and promoted the development of nuclear power technology and has supported, subsidized, and promoted the introduction of the technology by private enterprise. This has taken place despite the fact that the public will be exposed to very small increments of man-made radiation which, although never demonstrated to have resulted in manifest injury, are clearly regarded as at least undesirable.[11] The public is also called upon to assume the "exceedingly remote possibility" of a nuclear power reactor accident that could cause damages to health, life, and property of enormously catastrophic consequences, dwarfing by many orders of magnitude any other conceivable catastrophe that might result from a man-made cause. Protection of the health and safety of the public rests with a regulatory scheme which places reliances on "engineered safeguards," but, as we all know, man's engineering genius is far from infallible.[12] To further press this point, when the marketplace flashed a red light which would definitely deter private investment in the technology because of the enormous potential public liability (for which adequate insurance coverage was not available) in the event of the accident, the Government's response was to enact the Price-Anderson Act to remove this "roadblock." That Act superimposes upon $82 million of private liability insurance protection (the maximum available) a Government indemnity of $500 million and cuts off any further public liability in excess of $582 million.[13] This completely eliminates potential liability as a deterrent, since it is not possible that any firm could sustain one penny's worth of liability out of its own pocket. And, finally, it is candidly admitted that the Atomic Energy Commission regulatory program "exerts all effort which could reasonably be expected to ensure that there is no undue hazard to the public health and safety <u>while at the same time no crippling obstacle is placed in the way of development of [the] industry</u> . . . . [Emphasis added.][14]

There is, I believe, substantial reason to believe that the general pattern will be to entrust regulation of new government-sponsored technologies to the agencies that developed them and have an interest in their application.[15] Let me add, however, that I do not think separation of promotion and regulation into two separate agencies would be a panacea. There is also reason to believe that expert bodies, rather than the public itself, will make the decision as to what risks the public will be required to assume in exchange for what benefits. Experts will decide on how much sonic boom the public can stand; how much radiation it can tolerate in an exchange for what benefits; when it should rain, shine, blow or snow in the public interest; what kinds of children the genetic engineers should produce to further policy objectives; etc.

I believe the only effective mechanism for protecting the public against the onslaught of new government-sponsored technologies is the type of devil's advocate mechanism I have proposed. Indeed, such a mechanism is much more necessary in this case of government-sponsored technology because it is the only effective means for building restraints and deterrents into the system. Such an agency would give the public a full opportunity to determine whether it wants the government to develop a new technology which involves potential risks and, if so, the conditions under which development should occur. It would force full consideration of potential hazards upon the government at the developmental stage; and it would compel regulatory agencies to give greater weight to the public health, safety, and security than they do to the benefits of the technology. I cannot conceive of any other technology assessment approach which would not involve the cost of significant loss of democratic freedom in permitting an elite group of experts to determine the extent to which individual rights will be subordinated to the expert's conception of the public welfare in the name of technological progress.

Let me try to recapitulate very briefly. I am distrustful of the experts. Scientists and engineers have a bias in favor of accomplishing what they think can be accomplished. Their assumption that the problem of effective social control will take care of itself at an appropriate time is politically incorrect. In a government whose Executive and Legislative branches are committed to achieving the benefits of science and technology, excessive reliance is placed on the judgments of experts because of the unfounded myth that ordinary mortals are incapable of understanding the issues. What is necessary is that there be injected into the assessment process a clear and vigorous articulation of the negative factors in language comprehensible to the layman. This will compel the proponents of the technology to present their case in similar language, and the decisions will be made in the rough and tumble of the ordinary political process.

I am fully aware of the principal argument which will be made in opposition to this proposal. It will run as follows. If the negative factors are presented to the public, to the Congress, and to administrative agencies vigorously and in an unbalanced manner, undue apprehension will develop because the recipients of this information will not be able adequately to evaluate the negative factors and place them in perspective. As a consequence, scientific and technological advance will be unduly retarded.

My answer to this is twofold. First, the proponents of the technology have greater resources and at least equal access to the eyes and ears of the public, the Congress, and the administrative agencies.

Second, I can conceive of no reason why the public in a democracy should be forced to accept benefits it does not want, whatever the reason, rational or irrational.

NOTES

1. This point was effectively made by Harold L. Price, the Director of Regulation of the Atomic Energy Commission, in 1961. Responding to criticisms of the major role played by lawyers in the AEC's regulatory and licensing program, Price, a lawyer, said:

> "These questions are not deep scientific questions that can only be resolved back in somebody's laboratory. Sure, we need and could not move without the help of competent technical people, and we get that help. We who are not technically trained cannot determine the calculations, and we cannot determine what the technical safety question is, and, therefore, what the risk is. But once that has been identified to us, you and we and anybody with reasonable training can make a common-sense judgment as to whether the risk is acceptable or not."

2. This phenomenon is manifest in the flouridation controversy. For an example of the manner in which the establishment treated one dissenter who <u>was</u> permitted to testify before the Joint Committee on Atomic Energy in 1965, see U.S. Senate and House, Joint Committee on Atomic Energy, Hearings on Proposed Extension of AEC Indemnity Legislation, 89th Cong. 1st sess., pp. 155-76.

3. Special procedures would, of course, be necessary where classified information is involved.

4. The demand side of the market will indicate the extent to which buyers will pay a particular price for a product with hazards, the extent to which they will pay a higher price for the product with more safety built in, and the extent to which they may accept the hazards at a lower price.

5. For example, in 1931 the Wisconsin Supreme Court decided that a careful driver who splashes muddy water on a pedestrian on a rainy day should not be held liable to the pedestrian. The court balanced the benefits against the risks: "The benefit of allowing people to travel under such circumstances so far outweighs the

probable injuries to bystanders that such conduct is not disapproved." Osborne v. Montgomery, 203 Wis. 233, 234 NW 372 (1931). This type of balancing operation is performed without pretense of nice mathematical precision. Rather, it is a gross process based on the informed and experienced judgment of judges who reflect community values.

6. There is little direct evidence to support this proposition because, since a new technology when introduced is usually closely based upon and related to previous experience, insurance is generally available at reasonable rates. In the case of nuclear power, however, when the new technology represented a quantum jump over previous experience and previous levels of potential public liability, adequate liability insurance was not available at any price, and it is clear that there would have been no private investment in nuclear power in the absence of alternative arrangements to eliminate the threat of substantial uninsured liability. See below.

7. Admiral Rickover has described the pattern as follows:

> "Warnings of scientists are rejected as 'unproven' or 'exaggerated.' Later . . . the argument shifts to an attack on the legitimacy of any kind of protective legislation. Such legislation would violate basic liberties, it is claimed; it would establish government tyranny and subvert free democratic institutions. If all this is futile, and legislation is imminent, there will be urgent demands that it be postponed until 'more research' can be undertaken to establish the appositiveness of the proposed law."

"A Humanistic Technology," address before the British Association for the Advancement of Science, London (Oct. 27, 1965).

8. There is also an intermediate case: private development of a technology where related government-sponsored work makes a substantial contribution. The government contribution operates as a subsidy, tending to cause the market to flash a green light for development and introduction of the technology at an earlier date. The principal effect will be that adverse effects will be felt by more people to a greater extent at an earlier date, thereby heightening the need for effective governmental technology assessment.

9. It is characteristic of most government-sponsored technology that assessments take place in the context of consideration of a relatively small line item in a large budget or money bill. Only rarely

is the assessment made in terms of consideration of an isolated public policy issue on which full attention can be focused.

10. I use atomic energy as my principal example in this portion of the paper. I do so because it is the most advanced of the government-sponsored technologies with an immediate impact on the public, and because there is more relevant information available as to the benefits and risks than for any other government-sponsored technology. I believe, however, that the atomic energy experience is typical of the manner in which other government-sponsored technologies have been and will be assessed.

11. There is no evidence that exposure to low levels of radiation have produced manifest somatic injury; nor is there any evidence that there is a threshold of exposure below which no somatic injury will occur. It is, however, generally accepted that <u>any</u> radiation exposure produces undesirable genetic mutations. These considerations lead to the general rule enunciated by the Federal Radiation Council that "there should not be any man-made radiation exposure without the expectation of benefit resulting from such exposure." In actual practice, the amounts of industrial radiation to which workers and the public may be exposed are established by the National Committee on Radiation Protection on what is candidly stated to be a "philosophy of risk" or "calculated risk" basis as they assess benefits and risks. Despite the fact that the Chairman of the NCRP has stated that the setting of radiation protection standards is not "basically a scientific problem. . . . It is more a matter of philosophy, of morality, and of sheer wisdom," the people involved in the NCRP's standards-setting function are all specialists in the relevant scientific disciplines.

12. Because the technology is new and leapfrogs experience, the engineered safeguards have little basis in experience and reflect primarily the predictions of experts that they will be effective.

13. A 1956 Brookhaven National Laboratories (WASH-740, "Theoretical Possibilities and Consequences of Major Accidents in Large Nuclear Power Plants") report estimated, on pessimistic assumptions, that a serious power reactor accident might cause as much as $7 billion property damage in addition to substantial personal injury and loss of life. A more recent study, the details of which have been suppressed, led to the conclusion, as stated by the AEC Chairman, that because of additional experience the chances of such an accident are even more exceedingly remote, but because more recent power reactors are larger and are located closer to population centers, the damages that might result could be even more substantial.

14. AEC, Regulatory Review Panel Report (July 14, 1965).

15. The difference in approach between the typical administrative agency concerned only with the protection of the public interest and the new-fangled agencies which try to mix protection of the public interest with promotion of technology is vividly illustrated by the example of radiation preservation of food. For many years the atomic energy establishment has been developing a technology for using radiation to preserve and sterilize food. It believes the technology is now ripe for practical use by food processors, and it is clear that if AEC had regulatory jurisdiction the technology would now be used. Through a fluke in the law, however, radiation is a "chemical food additive" subject to regulation by the Food and Drug Administration. The FDA has reduced the atomic energy establishment to a state of petulant frustration by rejecting its claim that irradiation of food should be permitted because it has been tested with <u>no indications that irradiated foods are harmful</u>; the proper test, FDA insists, is that there be a <u>demonstration that it is safe</u>.

DISCUSSION OF THE ADVERSARY PROCESS IN TECHNOLOGY ASSESSMENT

Professor Green began the discussion with a review of the basic premises from which his argument proceeds. In his view, discussions of technology assessment tend to be imprecise because of a failure to separate distinct aspects of the problem before seeking solutions. One of the primary distinctions which must be made is that which exists between private technologies and the methods necessary for adequate assessment of such technologies on the one hand and government-sponsored technology and its assessment system on the other.

Of the two, Dr. Green is less worried about assessment of private technologies. For such technologies,

> at the most basic level we find that the marketplace provides a set of restraints which in themselves operate as a form of technology assessment. I think it is clear that the system works because there are competing interests involved; buyers and sellers have their own self-interests.

One element of the assessment system for private interests is the high degree of individual responsibility for hazards, which is imposed by the legal system through liability provisions. This compels a great deal of thought and analysis on the part of the developer of a technology before he is willing to embark on a new project. On top of the marketplace, a system of governmental assessment affects the private developer; "legislative and regulatory processes are, in a very real sense, a technology assessment." Government regulation involves a high degree of the adversary process. Dr Green noted that

> . . . the administrative agency which is entrusted with regulatory responsibility really doesn't function properly unless it is, in a very real sense, an adversary. It is an adversary in most cases because it is charged with the single responsibility of pursuing the public interest . . . as opposed to the private interest in exploiting and using a particular technology.

Thus, in the case of privately pursued technologies, mechanisms exist for effective assessment. Of course, these private and governmental institutions of assessment do not always function adequately. For the market system to operate with full effectiveness as an assessment mechanism there must be adequate knowledge on the part of the parties involved. That knowledge must be available before the effects of the technology are strongly felt. These conditions are not always met because of the difficulty of establishing a clear causal relationship between hazards and results. As for the governmental assessment, it is often ineffective due to "inertia and stagnation on the part of both legislative and regulatory bodies."

The problems of government-sponsored technologies are, in the speaker's view, more acute than those of the private sector. For one thing, government technology develops much more rapidly; this means that "to the extent that the technology is hazardous, more people suffer more injury before it is possible to take effective action" to restrain the technology. No real restraints exist other than what Dr. Green characterizes as "self-restraint." Decisions to proceed are made on a calculated risk basis, with determinations as to acceptable risk made "largely on the basis of judgments by scientists."

It is taken as axiomatic that benefits of technologies are always adequately articulated. It is the purpose of the technology assessment mechanism proposed in the paper to correct this imbalance by "ferreting out the risks." The mechanism would also question the "uncritical acceptance" of expert opinion and the assumption that "our legislative bodies and the public are incapable of comprehending the information which is relevant to decisions (with technological components)" and would seek to make the important issues understandable to all of the interested public. This would compel proponents of technological application to respond in kind.

Dr. Green concluded: "I visualize that this will stir people up, that it will get people apprehensive, and I think that this is useful."

He did not have to wait long to see his prophesy fulfilled. A seminar participant responded to the paper with surprising vehemence:

> I'm sure you've all heard the homily that it takes a carpenter to build a barn, but any jackass can kick it down. This arrogant, gratuitous, and insulting paper by Professor Green is representative of all the jackasses in the world. His certitude in the correctness of his position is unmatched since the dogmatists of the Middle Ages. His nihilism would undo a million years of evolution for--make no mistake about it--what Professor Green is proposing here is nothing less than the complete destruction of civilization.

Later, this participant admitted that his wording was perhaps too strong, but that a response of this sort was necessary in order to balance the extreme position in the paper.

This "intemperate attack" was criticized by another seminar participant who rejected the notion that any efforts toward the regulation of technology lead to the destruction of the civilization that now exists and stated that "I don't believe that the jackasses are all wrong. Occasionally a jackass stands up and kicks in the right direction."

There seemed, at this point in the discussion, to be some feeling among those present that the prepared paper might be a spoof or "a jesting probe to goad us into thinking," but most of the comments which follow were based on the assumption that it was meant seriously.

An agency official thought that the paper proposed, in essence, the "institutionalization of conservatism." He expressed some concern with the manner in which society seems to establish one set of evaluative criteria for new social technologies--the proposal in the paper is one example--and another set for new biological and physical technologies. He asked the participants "Are you willing to be as skeptical about new social technologies as you are about new physical technologies ?"

The same speaker further characterized the proposed assessment mechanism as the "institutionalization of reasons for fear." This worried him because "we all know that demagoguery is fueled by fear, and it seems to me that you've proposed here the ferreting out and publication of fear." Another agency official took issue with this point of view. "Is it not the ignorant who are stampeded by fear?" he asked. The proposal seeks to reduce ignorance and to increase rationality. Dr. Green echoed this point, noting that the institution he proposes is to be a "responsible" body and would not go out of its way to frighten people.

The discussion turned to a consideration of some of the specific assumptions and points of the paper. A university researcher questioned the assumption that benefits of technology are always articulated. Citing the example of the automobile manufacturers reaction to electronic car development, he claimed that "large and powerful economic interests very often serve as a brake, rather than as an accelerator, for technological change." A Congressional Aide agreed that very often technological advances require more effective articulation. Dr. Green conceded that there are cases in which industrial concerns choose, for some reason, to suppress innovations. His point was not to deny this but rather to emphasize that in those cases in which economic motivation exists, the benefits are always articulated. The problems raised by industries sometimes hindering innovation were, he felt, beyond the scope of the paper.

A professor could not clearly visualize the role of scientific experts in the proposed mechanism. Even this "negative agency" would be required to get its information from "the scientists and technologists in whom you've stated little reliance should be placed." How is the information to be used? "If you can't use their scientific judgments pro, why should you accept their scientific judgments con?" Dr. Green responded that scientists would play a very important role in the kind of technology assessment mechanism he proposes, but he insisted that scientists should not make decisions on problems of "philosophy, morality, and sheer wisdom."

An industry representative said that he could not agree that private commercial applications create less of a problem in technology assessment than do government-sponsored technologies. And an agency official challenged the assumption that the market mechanism is an effective assessment device, citing as examples the use of polyvinyl chloride containers and detergents as two failures of the market mechanism. Dr. Green pointed out that he had not claimed that the market was completely effective but felt that it was "useful." It fails to work effectively when it neglects the "externalities" such as long-range environmental effects.

A university researcher believed that the value of an institution of the type proposed in the paper was that it provides a place for the citizen to go when he feels threatened by some technological advance. Now one can go to Congress, but effective action doesn't always result unless a willing champion for a particular cause is found.

A Legislative Reference Service staff member noted that it would probably be quite difficult to find people who would be willing to work for a one-sided organization such as the proposed assessment institution. Dr. Green agreed with this point when he stated that "the kind of organization that I propose is not viable for the reason that you could never get anybody to work for it." He even noted that while he had not intended his paper as a spoof or a "put-on," he felt that it might, in reality, be a spoof because of the difficulty of staffing the assessment body. Still, he felt that it was useful to discuss the concept of such an organization.

A significant polarization in the attitudes expressed during the discussion was noted and criticized by two participants. One, a university researcher pointed out that acceleration of technology without restraint and "reverting back to the aboriginites" are not the only two alternatives open to society. He hoped that we "could encourage and stimulate technology" while keeping it responsive to public demands. The other, a professor, felt that the proposal of an agency of the sort being considered without a careful definition of "bad" effects on technology runs a risk of missing the "big middle." The "big middle" is composed of "those things produced by technology which government agencies legally can't do a thing about because they don't harm, don't do you any good--all they do is bleed your pocket." Unless the agency broadens its definition of negative effects to include this middle ground, it fails to protect the consumer.

An industry representative agreed with Dr. Green's faith in the man on the street as a decision-maker. The essence of democracy is the ability of the people to decide upon actions to be taken in the public interest. A Congressional Aide said, however, that the best you can expect form the public is that "they rise to the occasion once in a great while." The public would quickly tire of becoming involved in all technology assessments. But an agency official replied,

> I think you neglect the tremendous propensity of the American public for being engaged in a soap opera. There is no shortage of people who want to go out and belly-ache about any conceivable kind of bad thing that's going on.

Another participant, however, still felt that "the public has a limited capacity for indignation." He claimed that we can't expect all problems to be handled by public controversy and that in the long run new laws and institutions with clearly defined responsibility are necessary. Still, another speaker noted that in many cases the public seems to take no action at all in the face of potential hazards (such as polyvinyl chloride bottles). "Are you saying that public inaction is a choice in these things?" he asked.

An agency official sought to examine just how far one can or should rely on the public to make decisions in its own interest. The question which must be considered is similar to the one which asks whether a patient should direct a doctor on the form of medication he is to receive. Technology assessment is, in a sense, a professional endeavor and "a very important function of a professional ethic is . . . that it requires action in the interest of the client . . . whether or not the client agrees that the action is in his interest." Thus, for instance, we impose the requirement that all people be immunized against certain diseases without in each case consulting the person to be immunized. He asked Dr. Green whether he truly thought that the public should always be permitted to make decisions in its interests. For example, should the public be permitted to take action which could be destructive to itself? Dr. Green held that the example of a doctor's treatment of a patient was not really relevant to the question at hand and repeated his point that

> . . . the question is entirely one of knowledge, of legislative responsibility, and of the public's ability to respond meaningfully. It seems to me to be axiomatic that if one is talking about a question of balancing risks to the public against benefits to the public, it is the public itself which ought to be the one to decide whether it wants those benefits in exchange for the risks.

Some of those present were concerned about the location of an institution such as the one proposed here. A professor and a Congressional staff member thought that a true adversary process would be best located in the Congress. But a Congressional aide disagreed. Congress, he held, is apt to be slow in getting mobilized on most issues, and technology assessment, to be effective, often requires a rapid response. Further, Congressmen serve their constituents and often vote for programs which are locally beneficial even if there are detrimental aspects on a national scale. Another participant held that an effective technology assessment mechanism should be located near the President in order to best influence national policy. An agency official objected to any location within the governmental structure. An arm of Congress is subject to manipulation; an arm of the Executive is ultimately subject to pressures such as those which confronted the National Bureau of Standards during the battery additive controversy. He concluded that the proper location is "in the public domain." Others believed that effective technology assessment requires more than one location; the discussion of this point concluded without any real consensus.

Dr. Green asked those present whether there could be any conceivable objection to allowing the public to possess full information on all matters. A university researcher agreed that information availability was to be desired but questioned whether a mechanism whose sole function is the determination and dissemination of information concerning the detrimental aspects of technological applications really provided a means for giving the public all of the relevant information. To this Dr. Green responded that the public is certainly not fully informed now (although a Congressional staff member insisted that in this country we are better informed than those in "any other society on the face of the earth") and that his proposed institution would help to fill some of the gap.

A university researcher said that the objections which had been expressed concerning the "negative" assessment mechanism had perhaps been overstated:

> We probably need quite a multiplicity of new institutions and mechanisms and ways of dealing with this technology assessment. I don't see anything incompatible between an agency doing so-called objective detailed analyses looking at all sides of a question and another group which is trying to bring out certain of the negative aspects . . . and to present them in an effective way that everyone can understand.

He proposed experimenting with numerous types of social institutions in order to develop effective mechanisms for technology assessment.

It was a consensus of the participants that a comprehensive assessment system is necessary and that some sort of adversary process can play a role in such a system. A participant summed up the feelings of the group by noting that almost everyone present thought that an objective agency is needed somewhere in the system and that a smaller number felt that an "adversary institution" of the kind proposed by Dr. Green could also play an important role in a truly effective assessment system.

CHAPTER

5

THE MANAGEMENT OF TECHNOLOGY ASSESSMENT

Louis H. Mayo

INTRODUCTION

This is not the first time in history that various societies have been concerned about technological innovation and the dire impact, actual or prospective, of new applications on existing interests, institutions, and value schemes of such societies. Some think, however, that we are now faced with a technological intrusion of such magnitude that it threatens devastation not only to the environment but to our fundamental social values and institutions. We can no longer easily assume that technology is routinely absorbed into the social matrix and in turn modifies social attitudes and practices. In the view of some serious observers, the new technology may very well obliterate the social matrix. In a most provocative article in the *Saturday Review* Wilbur H. Ferry of the Center for the Study of Democratic Institutions posed the question: "Must We Rewrite the Constitution to Control Technology?" He came out with an affirmative answer. Mr. Ferry concluded his piece with a quote from Robert L. Heilbroner:

> . . . the coming generation will be the last generation to seize control over technology before technology has irreversibly seized control over it. A generation is not much time, but it is *some time* . . . .[1]

---

*Presented April 23, 1969.

One may or may not agree with Professor Heilbroner's blunt assessment, but few would disagree that the misuse of technology can produce swift and tragic harm. Such an assertion, however, does not move us far toward the design of technology control mechanisms which will be most compatible with our preferred social values. We must work from, or out of, the existing context of conditions, trends, and resource constraints, whether we like it or not. Surely, we would not wish to discard every element out of existing technology assessment structure, even though some drastic revision may urgently be in order. At least it would seem advisable to assure ourselves as to why and in what respects our present assessment arrangements are inadequate. In any event, we have little choice but to recognize and work from the base of existing social institutions, eliminating, modifying, or adding assessment mechanisms and processes as necessary for the development of a "controlled technological system, capable of producing benefits by design rather than by accident."[2]

The basic thrust of this paper is to analyze the frequently advanced notion of a "total problem approach" to technology assessment. A tentative suggestion is offered of one type of institutional arrangement which might produce a close approximation to the total problem assessment approach. Such an arrangement would supplement and coordinate the "existing technology assessment function"; it would not supplant it.

Several terms and expressions are used throughout this paper which require tentative definition. "Technology assessment function" is employed to refer to technology assessment activities in a general sense. "Technology assessment" refers to the identification of the effects (direct and derivative--immediate, intermediate, and long-term) and the evaluation of the social desirability or undesirability of such effects as related to particular technological applications. "Mechanism" [such as a Congressional committee or the Office of Science and Technology (OST) or a Special Study group] refers to the organizational entity conducting the assessment process. "Process" refers to the decisional-operational procedures (adversary process, panel discussion, research study) utilized by the assessment mechanism. A particular mechanism and its customary decisional process constitutes an "assessment subsystem." The term "assessment system" refers to all assessment subsystems which have relevance to a given technological application (fluoridation, aircraft noise abatement) or a pattern of similar applications (food and drug assessment system). "Social subsystem" is used to refer to the social interactions and social effects taken into account in the assessment of a given technological application. Normally, "assessment subsystems" for a given application will consider only limited interactions and impacts of the total social subsystem affected by such application.

THE TECHNOLOGY ASSESSMENT FUNCTION: A BRIEF OVERVIEW

If we accept as a tentative definition of technology assessment, the identification of the effects (direct and derivative--immediate, intermediate, and long-term) and the evaluation of the social desirability or undesirability of such effects as related to particular technological applications, one is probably safe in asserting that the decision process in every substantial technological project involves to some degree technology assessment.

While "technology assessment" seems to be an expression of recent origin, examination of the literature will disclose innumerable examples of technology assessment as defined above, even though most instances are obscured within other frames of discussion. The volumes of Singer on _A History of Technology_;[3] Derry and Williams, _A Short History of Technology_ (1961);[4] Kranzberg and Pursell, _Technology in Western Civilization_ (1967);[5] Bronowski and Mazlish, _The Western Intellectual Tradition_ (1960);[6] Morrison, _Men, Machines and Modern Times_ (1966);[7] and Dupree, _Science in the Federal Government_ (1957);[8] are rich sources of information.

From earliest times we can gain some perception of the interaction between developing technology and society and the various means employed to evaluate the social impact of applications. We are advised by R. J. Forbes in _Technology in Western Civilization_ that in ancient Mesopotamia "on occasion technical projects were submitted to the scrutiny and advice of learned bodies of priests who formed advisory boards."[9] The military has often been the social organization taking the initiative in the assessment and application of new technologies. Surely the major technological projects through history have involved tremendous efforts in planning, resource allocation, and implementation, which have inevitably required some evaluation of benefits and costs, whatever the value system might have been against which such impacts were measured. Societies have been concerned with the effects of technological applications on health and well-being, the physical environment, national prestige, the economy, and certainly on systems of belief and processes of thought.

Assessments are usually intertwined with an ongoing decision process which renders it difficult to make "clean" extractions of assessment from other phases of such processes. For descriptive purposes it may be useful to apply the following rough distinctions among assessments: "reactive" (evaluations of impact after the application has been made) "projective" (evaluations of probable impact prior to the application of the technology), and "reactive-projective" (evaluations made through time from early to advanced applications of developing technology).

There is an almost endless list of technology assessments on a "grand scale" for which massive commitments of resources have been required. One can easily imagine the magnitude of the effort in terms of debate, analysis, planning, and implementation which was involved in such engineering projects as the separate and elaborate aqueduct and sewer systems of the city of Rome, the extensive land-reclamation projects in the Netherlands during the late Middle Ages, and the construction of the Suez Canal in the 1860's. Several "assessments" were made of the technical feasibility as well as the probable political and economic implications of the Suez Canal.[10] It is of interest to note that the assessment and planning of a complete national transportation network was carried out in our early history. The scheme, though never enacted, was put forward in 1824 by the Corps of Topographical Engineers under the War Department. Professor Dupree states in Science in the Federal Government:

> Despite constitutional scruples, the Congress increasingly appropriated money for roads and harbor improvements. One offshoot of Monroe's straddling position on the constitutionality of internal improvements was the Survey Act of 1824, under which the Corps of Topographical Engineers made a comprehensive plan for canals between the Chesapeake and the Ohio, along the Atlantic seaboard, and for a road from Washington to New Orleans. This plan, the only one the government ever attempted to make for the country as a whole, required considerable technical competence, and had it been executed, would have required even more.[11]

Technological applications have almost always had their detriments as well as their benefits. This is particularly evident with respect to environmental encroachments. For example, in the 1500's "dwellers in the iron-working districts of England and Northern France were complaining bitterly of the shortage of fuel and demanding checks upon the activities of the ironmasters"[12] who were cutting down acres of surrounding forests to supply their fuel needs for smelting iron, thereby depriving the local people of both easily accessible firewood and recreational areas. In the mid-1800's, English citizens protested loudly over the noisy, smoky locomotives. Many land owners arranged for intermittent firing of guns across their ground to keep out railroad surveyors. "Parliament, exercising the right of eminent domain, eventually overcame these difficulties for the railroad companies, but only at a price: as a concession to objectors, a change was included in railway charters requiring that locomotives must not emit smoke."[13] And the public reaction to large steam carriages "brought forth in 1865 the famous Red Flag Act, which required a flagman on foot to precede each steam vehicle."[13] The manufacture of explosives such as nitroglycerine posed serious hazards. "In Britain the most satisfactory conditions for explosive manufacture were codified in the Explosives Act of 1875, and similar legislation was introduced in other countries to safeguard the workers in the industry."[14]

The experience of this country with reference to efforts to protect the health, safety, and general welfare of its citizens as well as the natural environment, is given detailed treatment in Dupree's Science in the Federal Government. Congress had been intimately concerned with an extremely broad range of matters that have in some substantial manner involved technology assessment, as for example:

1. In the early 1800's, Dr. James Smith of Baltimore proposed the free distribution of cowpox vaccine as a protection against smallpox. Congress in 1813 "passed a law naming him vaccine agent and giving him the privilege of using the mails without paying postage." In 1820 it was proposed that a "national vaccine institution" be chartered by Congress. But an unfortunate mistaken shipment of smallpox scabs instead of cowpox vaccine to North Carolina resulted in the inevitable spread of the disease. Subsequent objections by a North Carolina Congressman defeated efforts to provide Federal aid to the institution and led to the abandonment of the idea of free national distribution of the vaccine.[15]

2. Certainly one of the most interesting technology assessment incidents in our history involved steam boiler explosions. Between 1816 and 1848 "a total of 233 steamboat explosions had occured in which 2,563 persons had been killed and 2,097 injured, with property losses in excess of $3 million."[16] While it was not until 1852 that stringent and effective laws were enacted regulating boiler construction, operation and inspection, the Franklin Institute had researched the problem in 1836 and made recommendations at that time which

embodied most of the recommendations finally adopted in 1852.[17]

3. The rapid railroad expansion after the Civil War brought with it some public health problems. "Cattle brought from the southern states, though remaining healthy themselves, infected northern animals in large numbers with a disease that often proved fatal [Texas cattle fever]." Congress appropriated $15,000 for the Commissioner of Agriculture to investigate the problem of livestock disease, but the results seem to have been unsatisfactory. A similar problem arose later with the introduction of refrigerated ships for the exportation of American meat to Europe. Hog cholera and trichinosis were the result. In response to this "crisis" Congress created a new Bureau of Animal Industries to investigate, regulate, and prevent the transmission of livestock-carried diseases.[18]

4. In the early 1860's, Spencer F. Baird of the Smithsonian noticed the "great diminution in the numbers of the fish which furnish the summer food supply to the Coast" resulting from fishing in the breeding grounds. He insisted that Federal regulation was necessary and that it should be based on a research program to ascertain "not only the biology of each species of fish, but . . . the ecology of life in the ocean." The independent Fish Commission was created in 1871 and was changed to the Bureau of Fisheries of the Department of Commerce and Labor in 1903.[19]

5. In the 1850's the introduction of the English sparrow upset ecological balance in an unexpected way. Private scientific societies such as the American Ornithologists Union began to survey the nation's wildlife and assess ecological factors in wildlife distribution. The national scope of such problems brought in the Federal Government, which in 1886 established the Division of Economic Ornithology and Mammology in the Department of Agriculture.[20]

6. The conservation plans of John Wesley Powell of the Geological Survey in the 1880's with respect to irrigation and Western land settlement met great hostility inside and outside the Government. As Dupree states: "The irrigation work was dead, and with it Powell's great plan of orderly settlement based on facts of environment as determined by science."[21]

Perhaps the crux of the historical import of technology assessment, relevant to present purposes, is that in the Western countries there has been a strong presumption that the impact of scientific inquiry and technological advance is socially beneficial. From Francis Bacon on we seem to have accepted the "science is good in itself" notion. Certainly the scientific approach, however superficial, pervaded the outlook of the philosophers of the Enlightenment, that high point of belief in human rationality and the potential of man to perfect himself and society on earth. The Royal Society (chartered 1662) and similar organizations promoted the idea that "investment in science was an investment in prosperity."[22] Such organizations as the Lunar Society (1775-91) were more "practical minded" and socially sensitive to impact of science and technology than the more prestigious Royal Society.[23] Even the human wastage and misery inflicted by the early nineteenth century Industrial Revolution did not greatly diminish our infatuation with science and technology.

But many voices were raised over the abuses of expanding industrialism supported by technological development. Not only were alarms sounded of the degrading human situation resulting from industrialism, but in some instances rather far-sighted action was taken to curb such abuses.[24] Yet the overriding attitude continued to encourage technological development. This was particularly true in America during the nineteenth century, when resources were abundant, the population was dispersed, transportation needs were critical, and individual initiative was given the widest scope. According to the authors of a Short History of Technology the technological progress achieved by the end of the nineteenth century was prodigious. They state: "For technology exists to produce goods and services, and the Western World had marvelously increased its productivity, measured in terms of human labor." They continue: "by 1900 . . . the unskilled laborer in all the Western countries was earning more and had a shorter working day than the privileged class of skilled workers before the Industrial Revolution."[25] But there were some who expressed misgivings, such as Thomas Huxley:

> He came to Baltimore toward the end of the last century to say that he remained unimpressed by all the power, natural resources, knowledge and machinery that had so greatly extended man's competence over his physical environment. "The great issue," he went on, "about which hangs a true sublimity and the terror of overhanging fate is, what are you going to do with all these things?"[26]

Nevertheless, the industrialism supported by coal, steam, and a burst of inventiveness and motivated by the excitement of "progress" and personal gain reflected a social attitude raised to a Constitutional right through the doctrine of "freedom of contract."[27] Furthermore, many of the truly dedicated efforts to protect a broader and longer-term concept of the "public interest" were blunted or defeated by the reluctance of the Federal Government to encroach upon the traditional bounds of the state "police power" over health, safety, and general well-being.[28]

As some of the preceding comments have indicated, this nation has applied various assessment mechanisms since the early 1800's, although we have relied primarily on the "market" system for guiding and shaping the nature of new technological applications. There have been notable exceptions, however, as with the long agitation for improved public protection from adulterated foods and drugs.[29] Further, the telegraph represents an early instance of initial government support for the "demonstration" phase of a new technological application.[30] Governmental regulation has, in general, been gradual and piecemeal and, as in the case of transportation, usually evolved as a reaction to public demand for correction of specific and severe adverse effects of particular applications. The Judical, Legislative, and Executive branches have all been participants in this assessment function, as have numerous business, civic, and professional associations.

Although the Interstate Commerce Commission was established in 1887 and the first Pure Food and Drug Act was enacted in 1906, many of our more prominent technology-based regulatory agencies and statutory measures to control technological applications were not established until well in the twentieth century.[31] For the most part, these agencies represent reactive measures rather than prospective efforts to assure development of a new technology in the public interest. Even broadcasting was not brought under regulatory control until 1927, after frequency interference became intolerable.[32] The development of nuclear energy represents perhaps the most outstanding example of new technology whose development began under government supervision and for which a reasonable well-ordered assessment structure has been maintained.[33] The Marine Resources and Engineering Development Act of 1966[34] was a noteworthy example of the establishment of special assessment subsystems for a pattern of interrelated technological applications. This represented an effort to develop coordination among all relevant governmental agencies and to design an overall national organizational structure for the advancement of the marine sciences and the rational exploitation of ocean resources for both national and international social goals. Pursuant to this Act, the Council on Marine Resources and the Commission on the Marine Sciences, Engineering and Resources were organized to undertake these tasks.

Increasingly, since World War II, technological developments have been initiated and supported by the Government or through combined government and industry efforts or government-university arrangements. Such government-initiated and supported programs are normally subject to a formalized assessment system, which includes the following subsystems (mechanisms and review-decisional processes):

--One or more agencies or departments

--Federal Council for Science and Technology

--Office of Science and Technology

--President's Science Advisory Committee

--Bureau of the Budget

--Congressional authorizing committees

--Congressional appropriations committees

--Congressional oversight committees

--Originating agency or department once the program is in operation

--General Accounting Office

--Official ad hoc commissions or study groups

--Government contractors or grantees

--Continuing cycle of annual or periodic reviews of ongoing programs by the Bureau of the Budget and other of the aforementioned mechanisms.

The interaction among these subsystems even with respect to one area of technology can be extremely intricate. But the assessment is not likely to be limited to governmental entities. Assessments may also be made by the following:

Users and operators of the technology involved

Community groups affected by a technological application

Mass media commentators

Specialized journals in science and technology

Scholars in various journals

Professional association committees and study groups

The National Academy of Sciences and National Academy of Engineering

Independent policy and program analysis groups

Private standards-setting organizations

Etc.

FRAGMENTED ASSESSMENTS AND THE TOTAL PROBLEM APPROACH

In the Introduction of this chapter it was stated that the basic analytical concept to be examined would be that of a "total problem approach" to technology assessment. The Daddario Statement on technology assessment speaks of examining technological applications in terms of the "total social framework."[35] Senator Muskie in commenting on Senate Resolution 78 to establish a Senate Select Committee on Technology and the Human Environment uses the expression "total impact of scientific and technological change."[36] The Staff Report supporting the Muskie Resolution states in part:

> The Subcommittee expressed its conviction that the "big issue" of the future will be the ability of government and particularly the Congress, to see and to cope with each technological problem in its entirety, and to join the social sciences with the physical sciences and engineering to solve such problems.[37] [Emphasis added.]

Attention should be given to what this concept means--or might mean--in the context of the existing technology assessment function. Technology assessment must usually be considered within a time frame spanning years. Most technologies develop gradually. Hence, the social implications may not be immediately perceived. The social impacts become more evident through time. The means of evaluating the consequences may change through time. As applications increase and the technology affects a larger segment of participants, the larger the number of interest groups that become involved as operators or otherwise as receivers of benefits or absorbers of costs. As participants increase, the number of assessment subsystems (assessment mechanisms and assessment processes) increases. Such assessment subsystems will normally be concerned with one special aspect of the problem. Only a few will give attention to the larger social subsystem affected by the application or pattern of similar applications.

There are various reasons why assessments by particular subsystems are truncated. The authority to deal with given aspects of a problem may not be with the mechanism, or the subsystem may have a capability for dealing with only a special aspect of the problem, or the data may not be available for it to make a useful assessment at a given point in time even within its special area of competence. Even if the data are available to establish the "present state of the art," an apparent lack of feasible alternatives for advancing the analysis toward a practical solution of the problem may inhibit any genuine assessment effort. One of the obvious and difficult barriers to reliable "public interest" assessments is the bias, overt or hidden, of most assessment subsystems.[38] All of the essential elements for a fully adequate assessment (one analyzing all of the variable interactions within the total social subsystem) can be brought into focus only at certain points in time. Hence, assessments are almost necessarily incremental and cumulative, with only periodic complete assessments being made or even feasible.

A most useful study could be made on the strengths and weaknesses of various assessment subsystems. For example, a court case usually deals with an after-the-fact situation and declares rights and duties flowing from a technological application with respect to some highly restricted issue. It is basically reactive as contrasted with a prospective assessment. Initiation is not within the control of the court. The problem definition or scope of the assessment is largely determined by the issue in contention. Information is selected and limited by relevance to the specific issue. In other words, the range of alternatives that can be considered is extremely narrow.

Executive departments and agencies usually have their scope of authority and responsibilities spelled out by a statute. Such authority may be a narrow mandate for a special type of R&D with respect to a given technology. It may be reasonably broad, as where the mandate calls for the administration of a given technology, such as broadcast communications, in the "public interest." Further the rule-making and policy-declaring authority of regulatory agencies provide a projective dimension to the assessment function. Congress, it would seem, of all the permanent governmental entities, has the broadest authority and flexibility to make inquiries and assessments. The charter of special ad hoc study groups can vary from an inquiry into a specific issue to a "total problem assessment."

An illustrative sampling of assessments of the so-called aircraft noise problem may give some specificity to the foregoing remarks. Some aspect of this problem has been treated by the Judiciary, Congressional committees, the Executive Office of the President, Executive departments and agencies, local government units, private engine and airframe manufacturers, special ad hoc study groups, individual scholars, and many other assessment subsystems.

In the United States v. Causby,[39] military aircraft flying as low as eighty-three feet over a chicken farm of the plaintiff "caused the chickens to fly into the walls in fright, and destroy[ed] the use of the property as a chicken farm."[40] It was held that such diminution of property value was a "partial taking" which required compensation under the Fifth Amendment, and thereby clarified certain specific rights and liabilities.[41] Griggs v. Allegheny County,[42] involved a similar "taking" through noise from civil aircraft to or from a municipally owned civil airport, controlled by a government control tower. The court was faced with the narrow question of determining the party liable--here the airport controller.[43]

An extract from a report of the Senate Committee on Commerce sets out in rather explicit and succinct form the fragmentation of the assessment and decision function, stating that its immediate concern was with noise abatement at the source:

> The concern with aircraft noise has not, of course, been confined to agencies of government. The several segments of the aviation industry have participated over the years in numerous efforts looking toward solutions for noise and sonic boom problems. Engine and airframe manufacturers have maintained programs in noise research.
>
> This investment by the industry is representative of one of the avenues of approach to aircraft noise reduction, that is, the development of aircraft which generate less noise. Another approach to noise reduction is through the establishment of special flight operating techniques and procedures. The third principal control technique which merits serious consideration is the planning for land use in areas near airports so as to make such use compatible with aircraft operations. This is a matter largely within the province of state and local governments. While all these techniques must be thoroughly studied and employed, the first order of business is to stop

> the escalation of aircraft noise by imposing standards which require the full application of noise reduction technology.[44]

In its Notice of Proposed Rule Making on "Noise Standards: Aircraft Type Certification," the Federal Aviation Agency (FAA) undertook as its first order of business to concentrate

> on the aircraft that are most likely to raise aircraft noise levels in airport neighborhoods. These aircraft include subsonic transport category airplanes regardless of means of propulsion and subsonic turbojet powered airplanes regardless of category.[45]

The Committee on Environmental Quality of the Federal Council on Science and Technology stated in its report:

> The objective of the Task Force was not an exhaustive treatise on the technical aspects of the noise problem; but rather an exposition of its dimensions and a proposal for a more adequate Federal program directed toward elucidating the effects of noise on man and his environment and developing means for abating the problem.[46]

The comprehensive staff report[47] prepared for the Committee on Aeronautical and Space Sciences of the Senate by the Legislative Reference Service gave the noise problem very brief attention, referring primarily to National Aeronautics and Space Administration (NASA) responsibilities, but noted the then recently published Office of Science and Technology (OST) Report on "Alleviation of Jet Aircraft Noise Near Airports,"[48] and cited the broad scope and detailed treatment given the aircraft noise problem in the Report.

Two studies have undertaken to place the aircraft noise problem in a relatively broad social subsystem. A summary report of the Aeronautics and Space Engineering Board of the National Academy of Engineering[49] did not make _noise_ the focal point of the study, but did consider noise in its relationship to other major variables in the context of the general civil aviation program of the nation. And the Report of the OST Jet Aircraft Noise Panel,[48] provided a multidimensional perspective to the problem. This Report probably more closely approximates a "total problem assessment" than any of the others referred to above.

TABLE 1

AIRCRAFT NOISE CONTEXTUAL MATRIX

REPRESENTATIVE ANALYTICAL APPROACH TO: 1) Technology Assessment
2) Utilization of R&D Capabilities
3) Organization for Technological Innovation

NATIONAL PROBLEM AREAS	SPECIFIC PROBLEM CONTENT
(Illustrative)	(Specification of Social Sub-System)
A. Urban Affairs	A. Aircraft Noise
1. Transportation a) Rail b) Highways c) Waterways d) Airways e) Etc.	1. Civilian A-C 2. Source Abatement
2. Environmental Quality Control a) Air Pollution b) Water Pollution c) Radiation d) Aesthetic Debasement e) Noise f) Etc.	B. Social Sub-System 1. Initiating Event 2. Conditions and Trends 3. Goal-Objectives Public Interest 4. Participants Operating Affected 5. Perspectives of Participants 6. Capabilities of Participants 7. Practices of Participants 8. Other Relevant Characteristics
2.e) Noise 1) Construction 2) Automobiles 3) Aircraft 4) Etc.	

COMMENT

In order to analyze, for example, the probable degree to which any of the noted participants might contribute to the assessment of the aircraft noise problem or apply R&D capabilities or other capabilities to the abatement and control of the problem, accurate data must be secured and analyzed with respect to each of the participants, such as:

1) Perspectives (Objectives, Associations, Etc.)
2) Formal Authority - if any
3) Resources in general, including the extent of investment in the technology
4) Capabilities having specific application to aircraft noise abatement and control
5) Practices relevant to this problem
6) Proposed future activities relevant to the technology or to the noise problem

ASSESSMENT SYSTEM →
Assessment Sub-Systems Mechanisms

POLICY ANALYSIS – PROJECT PLANNING – PROGRAM IMPLEMENTATION	DOT/FAA	NASA	OST/FCST	BOB	HOUSE COMMITTEES	SENATE COMMITTEES	DOD	HEW	HUD	INTERIOR	CAB	JUDICIARY	NATIONAL TRANS. SAFETY BOARD	TRANSPORTATION AUTHORITIES	ENVIRONMENTAL CONTROL AUTHORITIES	LOCAL GOVERNMENT UNITS	FOREIGN NATIONS	AIRCRAFT ENGINE MANUFACTURERS	AIRLINES	FOREIGN AIRLINES	AIR TRANSPORT ASSOCIATION	PRIVATE AVIATION	AIRFRAME MANUFACTURERS	AIRPORT OPERATORS	FUEL PROCESSORS/DISTRIBUTORS	NAT. AIRCRAFT NOISE ABATEMENT COUN.	ADJACENT LAND USERS	CIVIC ASSOCIATIONS	NAS/NAE	SPECIAL STUDY GROUPS	PROFIT/NON-PROFIT PPP	UNIVERSITIES	SPECIAL ADVISERS	CRITICS--GENERALLY
(Illustrative)																																		
POLICY ANALYSIS																																		
Problem Context Conditions Objectives (Specific)																																		
Alternatives																																		
Model Construction																																		
Outcome Projection																																		
Impact Prediction																																		
Evaluation																																		
PROJECT PLANNING																																		
Program Definition																																		
R&D: Needed Completed In Progress Proposed																																		
Prototype Testing																																		
Simulation																																		
Experimental Operations																																		
Program Organization Government Private Other																																		
Resource Development Administration Funding Facilities Skills																																		
Assessment Mechanisms and Processes																																		
PROGRAM IMPLEMENTATION																																		
Program Authorization																																		
Program Initiation																																		
Program Operations and Management																																		
Program Monitoring																																		
Feedback Links to Preceding Phases																																		

SUB-PROBLEM CONTEXTS (Illustrative)

- Engine Design
- Airframe Design
- Flight Operations
  - Techniques
  - Patterns
- Noise Level v. Safety Precautions
- Formal Noise Standards
- Continuing Noise TAF
- Land Use Management
  - Analysis
  - Planning
  - Program Implementation
    - Zoning
    - Building Codes
    - Insulation
  - Incentives to Noise Abatement Measures
- Financing A-C Noise Abatement Programs
- Psychoacoustic R&D
- Educational Programs
- "Lead" Agency Designation
- Matching PPP Capabilities to A-C Noise Abatement Needs
- Aircraft Noise Litigation
- Military Aircraft and the A-C Noise Problem

A reasonably complete representation of the aircraft noise assessment system, including the various types of operations in the policy analysis, project planning, and program implementation continuum that might be assessed, and the range of subsystems making up the assessment system is provided by the Aircraft Noise Contextual Matrix (Table 1). One might start here with transportation or environmental pollution as the broad social subsystem involved. If we commence with environmental pollution, it is readily apparent that this area can be broken down into air pollution, water pollution, radiation exposure, aesthetic debasement, noise, and so forth. Taking noise as the primary problem, sources can be identified as surface transportation, construction, aircraft engines, etc. The source of aircraft engine noise can be reduced further to types of aircraft or by a division between civil and military aircraft. The alleviation of the problem may be viewed as one of reducing engine noise, devising noise abating take-off and landing techniques, or of land use management. These three primary noise abatement target areas can be reduced further into numerous subproblems. If we select civil aircraft, it can be noted that there are at least thirty-five to forty entities or assessment subsystems (mechanism plus the decisional processes of such mechanisms) which are relevant in the aircraft noise problem context. Each will have its own perspective (interest or objectives), special capability, and ongoing and planned projects. Now, who has the responsibility for the assessment of the total aircraft noise problem? The Senate Report referred to above states that "the Secretary [of the Department of Transportation (DOT)] is now the focal point for Government-wide activity in the field of aircraft noise abatement."[50] However, the Secretary clearly does not have authority over all aspects of aircraft noise abatement.

The import of the foregoing is that only infrequently, if ever, during the evolutionary period of a given application will there be a comprehensive social impact approach taken by one of the assessment subsystems. Even when a supposedly total social subsystem approach is taken (an effort is made to appraise all of the significant factor-variable interactions within the relevant social subsystem), it will necessarily be based upon the information available at that time. While this may be extremely useful for continuing program planning in connection with future applications of the given technology, as by identifying probable hazards which might be avoided, such assessment is still only a stage of a continuing process of assessment--at least until the technology has reached full utilization or the social problems created by the technology have been eliminated or minimized by an innovative "technological fix."[51] Hence, a time dimension must be incorporated in a workable approach to the technology assessment function. More will be said on this point later, but it might be noted that Congress first gave serious attention to the aircraft noise problem in 1959 (many years later than the courts), but it was not until 1968, after several interim hearings and special studies, that Congress apparently felt sufficiently confident to specifically authorize the Federal Aviation Agency (FAA) to establish maximum noise standards under new Section 611, Control and Abatement of Aircraft Noise and Sonic Boom, as an amendment to the Federal Aviation Act of 1968. Further, the assessment went primarily to the abatement of aircraft noise at the source rather than to the total aircraft noise social context. But the point here is that the assessment of a relatively low priority social problem such as aircraft noise can involve a variety of assessment subsystems, each having a different set of objectives, capabilities, practices, feasible alternatives, and forums (formal or informal) for making its contribution. Such assessment subsystems make up, by definition, the assessment system for the particular technological application. But the critical question is: Do they all intersect at sufficiently frequent intervals to combine effectively assessment outputs, and, even if they do, does this add up to a complete assessment of the impact of the given application on the social subsystem affected? Who has this responsibility for assuring a total problem assessment other than Congress? The OST? Do we need a new entity or a structure of new assessment arrangements for this purpose?

The point of our total problem assessment deficiency is well made by Karl D. Kryter in a recent article where he states:

> In view of the costs and commitments of aviation facilities involved in producing and operating the SST [supersonic transport], it would seem prudent for various governmental and scientific bodies, if not the general public itself, to examine closely the pertinent data from psychological and sociological research and their relation to arguments for and against the overland operation of the SST. The general unavailability of an integrated interpretation of the implications of the psychological, sociological, and acoustical research related to the acceptability of sonic booms to people has prompted the publication of this paper. In the last analysis the sonic boom is a psychological-sociological problem, and it would perhaps be regrettable if all relevant information, such as it is, from these scientific disciplines were not available and discussed in the practical context of the problem.[52] [Emphasis added.]

ADEQUACY IN TECHNOLOGY ASSESSMENT

Reference to the Provisional Schematic: Technology Assessment Function (Table 2) will show the basic operational phases of a particular assessment. The assessment is initiated by some requirement (formal or customary) or some nonregularized stimulus such as a "crisis" situation. The task of problem definition may be "given" or it may have to be formulated by assessment subsystem. The assessment function, of course, involves a continuous interaction among the various operational phases noted rather than a mechanical first phase, second phase, etc., procedure. Most assessments of a particular technological application involve periodic assessments through time, the same assessment subsystem often making several specific assessment inputs into the continuing assessment function.

In most assessment situations involving a problem of any complexity, the establishment of the "definition of the problem" or the "objective of the assessment" or the "context in which the problem is to be analyzed" or, in short, the social subsystem encompassing the social variable interactions and effects to be assessed, there will be interplay among all of the operational phases shown on the technology assessment function schematic. Starting with a given assessment subsystem and the institutional mechanism (Congressional committee, for example) which is undertaking the assessment, then the formulation of the problem definition will or may depend upon the initiating stimuli, the character (authority, capability, practices, etc.) of the assessing mechanism, the availability of data relevant to an analysis of the social subsystem to be assessed, the decisional procedures which the assessing mechanism is competent to employ, the possible follow-up actions which may result, and the various impacts of alternative actions (social effects--immediate or remote, direct or indirect--on various participants, societal values, and community institutions and practices).

There are various concepts of adequacy that might be applied. Approaches will differ with the perspectives of particular "adequacy evaluators." Two different approaches will be given brief attention here. First, one might accept the social subsystem or problem definition posited by the assessing mechanism and then measure adequacy by the extent to which a full analysis has been made of the social relationships, interactions, and effects within the social subsystem posited. In this approach, one is primarily concerned with the information selection and decisional procedure phases of the assessment. The second approach would begin with an analysis of the adequacy of the

TABLE 2

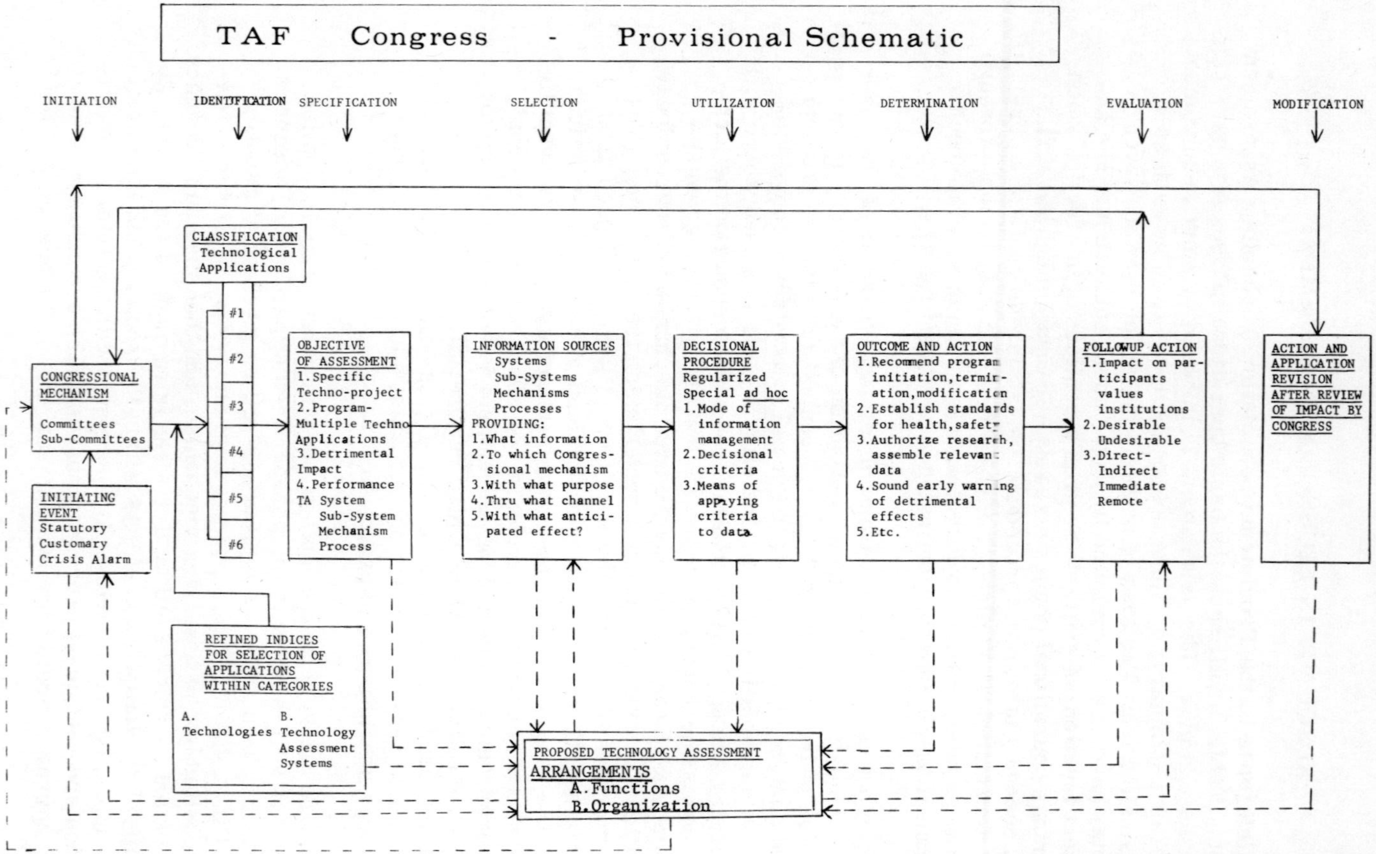

Question: How can the proposed Technology Assessment Arrangements improve the performance of each of these phases or operations in the Technology Assessment in terms of time, cost and reliability?

social subsystem posited. It would be necessary to take into account such factors as the general nature of the problem (social impact of a given technological application or applications), the urgency with which attention should be directed to certain adverse effects, the formal obligations and capability of the assessing mechanism, the availability of data relevant to various dimensions of the overall problem, and the probability of useful application of the assessment results to maximize social benefits and minimize social costs. The question would then be asked: Might a more comprehensive model (social subsystem) have optimized the utility of the assessment? Or conversely, might a less comprehensive model have been a more sensible "problem definition" in view of the factors mentioned above (such as capability of the assessing mechanism) or in view of the fact, in a given instance, that other assessment subsystems with high level capabilities had previously assessed or were currently assessing all of the relevant dimensions of a social subsystem coextensive with the "total problem" and the remaining need was only for a specialized assessment (within a constricted social subsystem in which the particular assessing mechanism has a recognized capability) in order to complete the full assessment of the "total problem"? In the <u>second</u> approach it is evident that the "adequacy evaluator" would have to make an analysis and evaluation of the information selection and decisional procedure phases in order to be assured that the information sources and the decision capabilities of the assessing mechanism can support a thorough analysis in the optimum social subsystem proposed.

It will be noted that both approaches to adequacy evaluation suggested that the <u>adequacy</u> of the overall assessment is largely dependent upon the <u>quality</u> of performance of the operations within the information selection and decisional procedure phases. The following criteria are relevant to measuring the performance of the operations within these two phases:

<u>Information Selection Phase</u>[53]

1. Availability and timeliness of data

2. Economy (cost of obtaining related to value)

3. Dependability (accuracy, reliability)

4. Comprehensiveness (contextuality, systematicness)

5. Openness (opportunity for participation)

Decisional Procedure Phase[54]

1. Problem definition

2. Specification of goals

3. Controlling contextual factors--recognition of

   Demands of participants

   Resources available

   Relevant institutional framework-legal process

   Customary practices

   Major movements and trends

4. Invention of alternative courses of action

5. Projection of more probable outcomes for each

6. Description-prediction of specific consequences of each alternative

7. Evaluation in terms of goals

The foregoing indices are to some extent objectively measurable. There are, of course, other criteria which would be applied in the evaluation of adequacy of assessment. Judgments will certainly be made on the "quality" of the analysis, as well as on the "operations" of the analysis. The skills of those performing the assessment and the methodologies used will also be matters for evaluation. Further, as given assessment subsystems are observed through time, reputations will emerge as to the quality level of performance.

One might say with considerable confidence that the problem of adequacy of assessment presents both a tremendous challenge and opportunity for conceptual and analytical study.

Previously, the fragmentation of the technology assessment function was mentioned in connection with the frequently stated need for total problem assessments. This is a fine notion, but the concept should be pursued a bit further. Assuming that "one-shot," total problem assessments are needed (which they are), it is not at all evident that such efforts are feasible with reference to certain applications at particular times. The assessment system simply may not have all of the necessary subsystems to produce the essential data, or the data may be available but there may still exist no mechanism within the assessment system for assembling and analyzing the full data input. Therefore, the alternative concept of a total problem assessment through time should be considered with outputs of the various subsystems being cranked into the continuing assessment as feasible. Such continuing approximations to a total problem assessment would be responsive to changing social demands and to new data developed on previously recognized and significant interactions in the social subsystem affected by the application. Further, the concept of total problem assessment through time provides the means for applying the sharpest possible focus to the selection of the social subsystem to be assessed. In other words, it provides an excellent means of testing the adequacy of an assessment in accord with adequacy approach #2 described above.

The evolution of the assessment function with respect to the highway-motor carrier social subsystem may serve as a useful device for illustrating adequacy approach #2. We are here concerned with the assessment function, not with the merits of the issue. During the last Congress, Senate Bill S.2658 brought to the focus of attention the prospect of Federal establishment of increased maximum sizes and weights of motor carriers operating in the Interstate Highway System. It is clear from the Report of the Senate Committee on Public Works that the assessment supporting the favorable recommendation of the bill was narrowly limited with primary focus on the physical capacity of highways to accomodate the larger vehicle. Under "Purpose of the Legislation" it is stated:

> S.2658 as reported with amendments will facilitate a more efficient and economic use of the Interstate System and ensure that the vehicles using that system will not unreasonably and unnecessarily impair its serviceability or durability.[55]

In making this recommendation the Committee was supported by the Bureau of the Budget and the Secretary of Transportation. In the letter of the Secretary of DOT to the Chairman of the Senate Committee it was noted that pursuant to the Federal-Aid Highway Act of 1956 the Secretary of Commerce had undertaken to "determine future maximum desirable dimensions and weights for vehicles operating on the Federal-aid highway systems"[56] and that a report had been made to Congress on August 18, 1964.[57]

The Committee Report also stated that "highway safety will not be jeopardized as a result of the proposed maximum allowable sizes and weights."[58] In support of this conclusion the Report stated that the "evidence presented . . . with regard to highway safety did not demonstrate a meaningful relationship between the sizes and weights under consideration and the incidence of traffic accidents."[59]

Nevertheless, even though some attention was given to highway safety and even to increased "user charges" in addition to increased maintenance and construction costs, one might view this assessment as inadequate in terms of a total problem approach to assessment. The overall social subsystem affected in some manner by this proposed legislation was hardly touched in this assessment. Why wasn't the motor carrier size and weight factor considered in a social subsystem, as for example:

1. A component of a national transportation "system" (all types of carriers for people and products) including an assessment of the impact (on participants, values, and institutions) of the interaction of the transportation subsystem with other major social subsystems

2. A component of a national <u>freight</u> transportation "system" including all modes of carriers with an assessment focused primarily on the implications for speedy and economical service to all relevant distributors and consumers

3. A component of the national highway program of primary and secondary roads, such subsystem including a consideration of the implications of the national highway program for <u>all</u> highway transportation activities and an analysis of the social, economic, and political impacts of such activities on other social subsystems (excluding, for instance, other modes of transportation)

4. A component of the special 41,000 mile Interstate Highway only--otherwise as in #3 above.

However, in a letter from the Secretary of Commerce of August 18, 1964, to the Speaker of the House the point was made that such a proposal should be considered as only a phase in a continuing process of <u>progressive implementation</u>[60] in adapting vehicular standards. Therefore, it would seem prudent to appraise the assessment function with respect to highway-motor carrier applications in the time dimension of our developing highway program. We might start in 1956, when Congress enacted major Federal highway aid legislation which reflected the request of President Eisenhower for

> [A] grand plan for a properly articulated (highway) system that solves the problems of speedy, safe, transcontinental travel, intercity transportation, access highways, farm-to-market movement, metropolitan area congestion, bottlenecks and parking.[61]

A special Presidential Advisory Committee, of which General Lucius Clay was Chairman, had submitted a report which the President attached to his Message to Congress on "A Ten-Year National Highway Program." While the President's Message directed attention to the "nation's highway system" rather than to a social subsystem including all modes of transportation, the social subsystem posited by both the Advisory Committee Study and the Congress clearly showed the intention to include all significant social interactions and effects of this proposed "national highway system."[62]

Analysis of the Congressional Committee Report shows that an extremely wide range of engineering, financial, and social factors was considered.[63] From our present perspective, however, we note that some factors were given relatively little consideration and that others were given no attention whatsoever. The Advisory Committee and the Congress seemed to be much more concerned with matters bearing upon the efficient implementation of the highway program rather than with cumulative and qualitative social impacts, particularly those which might be detrimental. One might feel, of course, that the exclusion of other modes of transportation was sufficient to disqualify the assessment supporting this legislation as a total problem assessment. The Clay Report did explicitly state, however, that

> This Committee was created to consider the highway network, and other media of transportation do not fall within its province. This relationship between the several modes of transportation is under study by other Government agencies and special committees fully informed of these views.[64]

Accepting this limitation, however, it is still to be noted that no consideration was given to environmental pollution, which would increase due to the growing traffic volume: air pollution from exhausts, engine noise, resulting aesthetic debasement, and the derivative health hazards from the foregoing sources would all result. Nor was a great deal of attention given to the relationship between the increased number and size of motor freight carriers and the possible increased hazards to private auto drivers and passengers. Further, there were several factors which the Congress could not treat adequately for lack of relevant information.[65] In view of these gaps, must one conclude that the assessment was inadequate? A negative conclusion without further analysis is not necessarily warranted. In the 1956 social context of urgent highway transport needs, coupled with the relatively low priority then given to certain social values which now have been drastically upgraded, one might reasonably conclude that Congress posited an extraordinary broad social subsystem for assessment. But whatever view one might take of the 1956 assessment, our primary concern here is with the scope of the assessment that should have been made of the proposed increase in motor carrier size and weight as set forth by Senate Bill S.2658 in the last Congress. One might find it difficult to justify the extremely narrow scope of interaction (social subsystem) considered in connection with S.2658; limited, as noted, primarily to an inquiry of the ability of the highway network to withstand the wear and tear of the larger motor carriers. But if we view this assessment in the sequence of assessments made by Congress between 1956 and 1968, we may get a different perspective. Taking the assessment supporting the 1956 national highway network legislation as the basic assessment, the following legislation is illustrative of Congressional action (and assessments) during the interim period:

Federal Highways, Billboard Regulations (1958), further legislation (1959), (1961), and (1963)[66]

Air pollution from motor vehicle exhaust fumes (1960)[67]

Clean Air Act (1963)[68]

Motor Vehicle Air Pollution Control Act (1965)[69]

Air Quality Act (1967)[70]

Interstate compacts on traffic safety (1958)[71]

Specifications for hydraulic brake fluids (1962)[72]

Minimum standards for seat belts (1963)[73]

Provision for State Highway safety programs (1965)[74]

National Traffic and Motor Vehicle Safety Act (1966)[75]

Highway Safety Act (1966)[76]

Standards for bridge inspection (1968)[77]

Highway Beautification Act (1965), additional legislation (1968)[78]

Establishment of the Department of Transportation, including

the National Transportation Safety Board (1966).[79]

Further action relevant to the National Highway Program includes provision for broader community participation in highway planning.[80]

In view of the manner in which Congress has approached the highway-motor carrier issue (and perhaps, must approach the problem, that is, in terms of specific bills), can it be plausibly maintained that Congress has been in effect making an organized total assessment on a cumulative, sequential basis? Has consideration by Congress of the various aspects of the Highway Program been "programmed" in such manner as to assure the closest possible approximation to a total problem assessment through a reasonable period of time? To put the matter more explicitly: Was there any need for a broader social subsystem to be considered by Congress than the Senate posited in connection with S. 2658? Was it not a wholly adequate supplementary assessment in the context of previous related assessments? After this bill breezed through the Senate, it ran into heavy opposition in the House. The Automobile Association of America launched a national advertising campaign against House passage.[81] The treatise of Professor Fuller of Washington State University was inserted in the <u>Congressional Record</u> for the apparent purpose of inviting attention to a much broader scope of social effects, primarily "social costs," than had been given attention by the Senate.[82] In short, some people clearly felt that the scope of the social subsystem posited by the Senate Report was not sufficiently comprehensive to assess properly the significant social impacts of this proposed legislation. But what should the scope have been in light of the many previous Congressional assessments relating to highway-motor carrier activity?

Several questions emerge from this recital of the legislative history of S. 2658 which seem highly relevant to the improvement of the technology assessment function. In continuously developing programs involving substantial scientific and technological applications, how does one determine the <u>optimum social subsystem</u> to posit for examination at a given point in time? To what extent does the optimum social subsystem depend upon the nature of the <u>initiating</u> or <u>carrier issue</u> (or proposal such as S. 2658), which invites or compels the positing of the social subsystem to be examined? Surely, every time some aspect of highway-motor carrier activity needs attention, it is not feasible nor necessary to conduct a total assessment involving the consideration of every conceivable decisional factor and social impact (direct, indirect, immediate, and remote) that might flow from such action. There is hardly a limit to the scope of social impacts that can be investigated. What kind of assistance might be useful to the relevant Congressional committees in determining the "scope of the social subsystem" to be evaluated prior to action on such a proposal? If the specific bill itself sets the limits of the inquiry, then we in effect beg the question we are supposed to be exploring and thus fail to advance beyond our current level of fragmented, unrelated, and inadequate assessments. Should Congressional committees request from the responsible agency or agencies explicit and systematically organized suggestions as to the scope of the social interactions that should be explored in connection with specific legislation, taking into account previously developed data from assessments from all relevant sources? Is there a continuing and critical question of positing the optimum social subsystem to be examined in a given Congressional hearing assessment? Or do organized community interest groups in conjunction with comments of relevant Government agencies assure that an adequate scope of social interactions and impacts is considered?

Since we have a time dimension to consider and an assessment system composed of a diversity of assessment subsystems, it would seem that we might accept as one provisional model of the assessment function the notion of an aggregative approach to total problem assessments (total problem assessments through time). This is roughly what the Congressional model of assessments amounts to. It will at intervals bring together the various assessment subsystems to a greater or lesser degree. Whether such periodic efforts do in fact bring together all of the relevant subsystem outputs is, of course, another question. Even if so, there is still the further question of whether the combined outputs are utilized in such a manner as to extract a complete social subsystem assessment as of that time. The final question is whether in many assessment patterns the full matrix of relevant assessment subsystems, even if properly utilized, add up in their complete contribution to what amounts to a total problem assessment of the applicable technology.

Periodic study groups or standing intergovernmental committees or even Congressional subcommittees very seldom assume the responsibility for making a total problem assessment with any degree of thoroughness. Even if this is the objective, limitations on authority, time, information, or other essential resources preclude a wholly satisfactory total problem assessment, including such elements as what scope of the social subsystem should be treated, what has been done which contributes to the analysis of such a social subsystem, and what still needs to be done in terms of policy analysis, project planning (including research and development), and program implementation. Surely we need a more systematic and reliable mode of managing this kind of inquiry.

What would seem indispensable to our getting into a position of "confident control" over such social subsystems, as for example, that encompassing the aircraft noise problem, is an activity which will construct a comprehensive matrix of the total social subsystem in which the aircraft noise problem is located, and maintain a continuing appraisal of the assessment status of the problem. Such a matrix function would necessarily include the following:

1. Identification of the participants (operational and affected) within the total social subsystem

2. Characteristics of the various participants (which are to be considered as assessment subsystems)

   a. Perspectives (demands, associations, etc.)

   b. Formal authority (if any)

   c. Resources--general

   d. Capabilities having specific application to aircraft noise problems

   e. Practices relevant to this problem

   f. Proposed future activities relevant to this problem

   g. Etc.

3. The relationship of the activities of particular participants to the various components of the policy analysis, project planning, program implementation continuum, noted in the aircraft noise contextual matrix

4. Particularization of the following:

   a. Significant controlling conditions and trends in the social subsystem posited

b. Present and prospective social impact of existing or prospective practices related to the problem context

c. Research and development done, in progress, or proposed

d. Governmental regulatory measures taken or proposed

e. Alternative means available or in prospect for aircraft noise abatement (whether with reference to source or affected participants and activities)

5. Ways in which the total social subsystem posited can be reduced to more limited social subsystems for purposes of treatment by particular assessment subsystems

6. Modes of coordinating specific assessment outcomes so that at appropriate intervals the full effort can be combined into the equivalent of an assessment of the total subsystem posited.

The mechanics of constructing, maintaining, and monitoring the matrix function is open, but clearly all of the new analytical, communication, and managerial concepts, as well as automatic data processing and simulation techniques, can and should be utilized. The resultant would be the establishment of means for effectively relating capabilities to needs. The aircraft noise contextual matrix is a rudimentary representation of what would be involved in this task.

It would seem that the OST Panel Report on "Alleviation of Jet Aircraft Noise Near Airports"[83] reflects one of the closer approximations to the comprehensive matrix approach. But even the enlightened effort of the OST Panel should be appraised for its limitations in terms of the comprehensive matrix function. First, the Report represents an ad hoc rather than a continuing effort. Further, while the range of content is remarkable, one is left, nevertheless, with some uneasiness as to whether all of the major significant social interactions and relevant assessment subsystems were identified and their perspectives, capabilities, practices, and prospective future contributions fully appraised. Perhaps the critical question, for purposes of analyzing the technology assessment function, is why this special ad hoc group had to be set up in the first place. Does the persistent practice of either establishing or recommending the appointment of special assessment subsystems (groups, task forces, committees, commissions, and so forth) provide most persuasive evidence that a much greater degree of "regularity" needs to be brought into the technology assessment function?

TOWARD STRENGTHENING THE TECHNOLOGY ASSESSMENT FUNCTION

While recognizing that attempts to give systematic and comprehensive treatment to the technology assessment function are of fairly recent origin and that current study projects are yet to report, a few tentative notions are suggested below which might usefully be given some critical attention.

Initially, we can probably agree that several identifiable conditions have and continue to contribute to the inadequacy of present assessments:

1. Only recent emergence of interest in developing effective appraisal systems for social system analysis as contrasted, for example, with economic market analysis

2. A strong disposition to react to crises rather than to anticipate outcomes and evaluate consequences in terms of alternative schemes of social indicators

3. The fragmentation of technology assessment subsystems; the truncation of assessments made and the concomitant lack of management and informational systems for assuring total problem assessments

4. The lack of conceptual thinking on the technology assessment function, including alternative notions of adequacy of assessment, which reflects current deficiencies in the professionalization of the function

5. The failure to utilize the resources (methodology, techniques and professional skills) for technology assessment we now possess, including the lack of support for the level of effort required to assure the effectiveness of such utilization

6. An inability to translate intellectual discussions or analytical findings of future detrimental consequences of given technological applications into a public perception that a problem exists

7. Alarm on the part of some segments of society that undue emphasis on technology assessment will inhibit technological innovation and hence limit the range of individual opportunities and social progress.

The previous discussion has stressed that the technology assessment function is both pervasive and fragmented. One would hardly expect the situation to be otherwise in our pluralistic society. Furthermore, the magnitude of this function is enormous. This fragmentation reflects many conditions that work against the total problem assessment approach: continuing specialization of knowledge and skills, highly compartmentalized jurisdictions and authorities, narrow assessment foci, and so forth. Since many of our traditional values are supportive of divisiveness in assessment activities, progress toward the effective management of total problem assessment will require strong and substantial efforts. Emphasis must be given to new professional skills, institutional arrangements, and value reorientation to some extent.

For the general support of a more adequate technology assessment function there is need for strong encouragement of greater professionalism in this activity. Efforts in the universities toward the development of problem-oriented, policy analysis professionals are rather limited. While some progress is being made, the traditional disciplinary boundaries and the surge toward greater specialization makes it exceedingly difficult to introduce a total problem perspective. Concepts of "systems thinking" and techniques of information and simulation management do, however, provide new and effective means by which those disposed toward problem-oriented approaches can utilize specialized knowledge for policy analysis. Further, there is clear evidence than an increasing number of young professionals are interested in breaking out of the constrictions of a traditional discipline or profession in order to deal with social problems in a broader context.

One of the more critical deficiencies is the lack of a satisfactory information system for dealing with the appraisal of public programs, including major projects having significant technological components. There is much in the existing literature bearing on technology assessment: court decisions, regulatory agency hearings on rule-making and policy positions, special task group reports, studies of the National Academy of Sciences and Engineering, scholarly studies, industry R&D reports, and so forth. But this is not well-organized for effective use in terms of the problem approach. It is an extremely time-consuming and inefficient process to bring together relevant and important work that has been produced in the past. Many significant reports of special commissions and committees established to assess a techno-social problem context are almost impossible to obtain.

The materials could be set out in various ways: digests of case studies, reports of relevant entities, Congressional hearings, agency appraisals, etc. But the emphasis must be placed on a problem approach rather than a rule approach, as in the legal references. In other words, the organization should emphasize alternative thinking rather than rule-deductive thinking. We must have materials which will enable us to compare and contrast problem definitions, social goals, descriptions of existing conditions, influential trends, available resources, alternative means of achieving objectives, and the predicted impact of alternative means on various schemes of social values. The establishment of such an information management system would be a major step forward in a systematic approach to technology assessment. It should be designed both to support systematic inquiry into techno-social problems and to utilize fully the concepts and analytical tools of the new technology such as systems analysis, automatic data processing, and advanced techniques of simulation.

Other conditions will require some modification. Economic market assessments are no longer suitable for many technological projects. It is therefore essential that forums and decisional arenas be provided wherein the full range of social values can be expressed and alternative courses of action can be judged on the basis of the fullest information and analysis it is possible to obtain. So long as the predicted consequences of technological applications differ among the "experts," as will surely be the case, and so long as value priorities differ among various segments of the community, there will obviously be differences of opinion as to the feasibility and desirability of technological applications. Whether early warning should be made of possible beneficial or adverse effects will be debatable with reference to many applications. Provisions should be made for a vigorous "adversary process" in the broadest and most legitimate sense. Confrontation should be expected in providing alternative views of social objectives, problem definition, existing conditions, influential trends, courses of action, and the probable consequences flowing from each such course. The "open forum" approach is the means of soliciting the broadest community participation and assuring that the full assessment resources of the nation are drawn upon.

There is need to assure a continuous upgrading of the performance of technology assessment systems, to see that new systems or subsystems are initiated when necessary to cope with new or projected technological applications, to see that useless duplication is eliminated where feasible, and to see that distorted and irrelevant outputs of subsystems are noted and dismissed. In this task, criticism of poor performance and persuasion to improve performance will probably be a better technique than any type of formal control.

A highly fragmented assessment function, however, has the inherent characteristic of rendering it extremely difficult to assure that the essential data and analyses are available at appropriate times, periodically or through socially permissable time spans, for a total problem assessment or an optimum assessment (which may not be a total one) of a given technological application. Hence, a coordinating or programming function is needed which is lacking in many techno-social problem contexts today. Reference to the aircraft noise contextual matrix should demonstrate the difficulties in making a total problem assessment even with respect to this relatively low-priority social problem (as compared to other national problem areas). No entity now has the authority (in view of Federal-state-local-private divisions of power and responsibility in this problem context) to assure a total problem assessment.[84] Congress has been reluctant to intrude into traditional areas of state powers, as is vividly illustrated by the treatment of aircraft noise and the various problems associated with the continuing development of highway-motor carrier technology. Hence, a total problem assessment requires persuasion and appeals to mutual interest in order to obtain the essential cooperation and support from state and local government entities as well as private sector organizations in the task of assembling all relevant information.

Pursuant to this point there may be a need for <u>interim organizing mechanisms</u> where it is not clear which existing agency has the ultimate responsibility for the initial assessments of a developing technological application. In this way Congress would have better means than it now has for securing the best assessment feasible at any given time.

One matter which must be given attention is the relationship between technology assessment and the follow-up decision process. The two are inextricably related. Most technology assessment subsystems are decisions. Technology assessment (looking at alternative ways of applying technologies to determine what the impacts will be in terms of social goals) is a part of every consideration of a new technological program unless the decision process is totally lacking in rationality. Our basic interest is in assuring adequate assessments upon which the socio-political decision can be made. It is therefore important that all public and private entities (assessment subsystems) and all segments of society affected by the probable action participate in the decision process. With certain exceptions, as with some Presidential commissions having a broad charter to inquire into a given problem, Congressional committee hearings now provide the most satisfactory forum for approximating a total problem assessment. Usually, though not always, most of the governmental, industry, and community entities having relevance to the problem either as direct data input assessment subsystems or as segments of the affected public have an opportunity to participate. The contribution of such participants is normally, by virtue of the nature of the technology assessment function, truncated and directed to a special aspect of the problem context. Special interests of one kind or another are advanced as should be expected. Even the particular committee may be a promoter of certain types of technological projects. Assuming, however, that the outcome in deciding what action to take is based upon the resultant of the conflicting views advanced, one must ask whether this procedure provides for a fully adequate assessment. We have noted that the assessment system for any particular technological application is made up of subsystems, each with differing objectives, resources, capabilities, and projects, and can supply only a partial assessment of the total problem. Since so many of the techno-social problems with which we must deal today are extremely broad in their social reaches, is it not absolutely essential that we recognize an additional dimension in the concept of an adequate assessment?

Congress, as has been noted, is at the apex of the technology assessment function. Congress is, in general, less constricted in its capability to approximate a total problem assessment than is the Judiciary, the Executive, the regulatory agencies, private assessment entities, the economic market, or even ad hoc commissions established for the explicit purpose of making a total assessment. Congress has the ability to extract or compel almost a full spectrum of informational inputs, despite such occasional obstacles as Executive Privilege. In this connection, it has persuasive powers beyond strict Constitutional authority. It is the ultimate authorizing, funding, and oversight branch, a fact hardly necessary to mention. Congress translates assessments into action-projects and programs. But the question remains as to whether information and analysis of the comprehensiveness and reliability needed for a total problem assessment of many techno-social problems is now available. It seems doubtful.

Perhpas through greater effort to systematize the continuing oversight of the adequacy of assessment systems Congress could improve the quality of the data it receives and acts upon. Congress could probably obtain additional assistance from the Executive and regulatory agencies if it demanded more comprehensive submissions of data, including the range of interactions and social impacts that should be examined in connection with any new proposal. At the least, response to this demand would be interesting. But in fairness, it would be necessary to provide many agencies with resources for policy analysis they do not now possess. Even so, while improved in-house assessments might be helpful, there would seem to be an imperative need for a separate structure of assessment mechanisms and processes (assessment subsystems) which would directly support the total problem assessment approach. This notion has been referred to above as adding a new dimension to the "public interest," one independent of any special, partisan interest. The assumption made is that there may be a "public interest" that, in some instances, will be something more or different from the resultant of the usual partisan interest conflict. At least, this is an assumption which might be explored. Some assessment subsystems now make a significant contribution of this nature, as for example the Science Policy Research Division, Legislative Reference Service, of the Library of Congress.

But again, the technology assessment function in this country is vast and complex. It seems unlikely that one monolithic evaluator, however well-supported, can do the job. Such an arrangement would obviously have many other disadvantages. One might think of a structure of several entities, or assessment centers, possibly but not necessarily being associated with other community institutions. Wherever located or however financed, they would have several common and indispensable characteristics: They would be independent, devoted to the enlightenment process--not to power. Their ultimate responsibility would be to the public, even though their assessments might be utilized primarily by legislative bodies, particularly the Congress. With respect to those patterns of techno-social problems each might undertake to assess, the approach would be in terms of total problem assessments. They would also monitor and appraise the adequacy of relevant assessment systems. They would develop the data banks essential to this task. They would be neutral forums for soliciting the fullest range of relevant views in the selected problem areas. They would report fully on the basis of assessments made in terms of information sources, assumptions made, and identifiable decisional components, so that such assessments could be critically and systematically compared with other assessments. They would not be merely an "on-call" activity, subject to any and every agency's demand for an instant response to a request. A basic purpose of such an institutional arrangement would be, however, to anticipate emerging problem areas so that useful assessment information could be supplied in a timely and orderly fashion.

Clearly, there would be some difficult problems relative to the operational procedures. What would be the ground rules for obtaining essential material if classified by a government agency? What procedures might be developed to obtain relevant information from resisting private sources? What information might be relevant but (for reason of other strong social purposes such as encouragement of economic competition) not available to such entities? Who would support such centers? Would it not be in the public interest for Congress to do so even at the risk of occasional embarrassment to particular committees or to other agencies of government? But support should probably be diversified if possible. In any event, it will be necessary that substantial support be supplied to perform the job properly. This must be a continuing function. Further, a long learning curve is to be expected, and it will take time to organize the resources and engage the quality of professionals who can cope with this type of task and who are dedicated to the importance of the function.

But Congress cannot perform the entire technology assessment function alone. What Congress does need is competent, comprehensive, and continuously current assessment data upon which it can confidently take action with reference to programs involving particular technological applications. The suggested structure, or a similar one, might very well provide the means by which the Congress can be confident that it is continuously aware of the status of the many dimensions of the technology assessment function.

A group of assessment centers stressing the total problem approach to assessment would provide the type of support essential both to the needs of Congress and to the general public. The assessment centers would assume the major burden for data collection and organizational analysis, relieving Congress of the necessity of having to resort to limited staff resources for this purpose. Instead of having to rely upon the relatively haphazard procedure of picking up bits and pieces of hopefully relevant assessment information on particular applications, or at best, having to rely upon the partial assessments of a fragmented assessment system, the Congress would have an orderly way of keeping informed on all significant techno-social problem contexts, existing and emerging. The loci for total problem assessment would be fixed. While the various centers would cover somewhat different areas of the technology assessment function, Congress would know which one was giving attention to what assessment system (meaning to which technological applications, existing and prospective).

Such centers could not only provide timely and reliable inputs into the final Congressional assessments, but also would be continuously sensitive to signals of impending or prospective severe detrimental impacts of technological applications. The assessment centers would contribute to continuity of the assessment function with reference to particular techno-social problem contexts. With such continuity should come a higher degree of reliability, less waste in duplication of effort, and a higher level of confidence in the assessments made. The adequacy of performance of the assessment systems for the more significant technological applications would be appraised; deficiencies would be pointed out in the assessment subsystems--a lack of assessment mechanisms or weaknesses in the evaluative processes of such assessment mechanisms. The centers could identify the necessary outputs needed from various assessment subsystems before a total problem assessment would be feasible. They could bring to the attention of Congress the need for timely consideration of given applications instead of leaving such assessment efforts to chance to the challenge of one or more influential participants, to the necessity of confronting "crises" where technology has "run wild," or to mere routine statutory or customary responsibility for oversight. The centers could further indicate the scope of the social subsystem that could optimally be considered and the range of relevant social interactions and effects. Or, in instances of prospective technological applications where a regularized assessment system does not exist, the centers could recommend either the establishment of a new entity or an ad hoc commission as an initial approach to a total problem assessment.

In any event, there is no escaping the need for the foregoing assessment tasks if the Congress is indeed concerned about the adequacy of the national technology assessment function. It appears highly unlikely that we shall ever gain confident control over technological development and technological applications until we have the means by which the closest possible approximations to total problem assessments can be made. But this will remain an elusive goal unless certain entities are given the <u>responsibility</u> for total problem assessments (including <u>formal authority</u> insofar as it is essential to the assessment function) and the <u>resources</u> necessary to adequately perform this enormous task of assessment information management.

CONCLUDING REMARKS

The diversity of approaches that may be taken to the analysis of the technology assessment function and the multiplicity of alternative arrangements which may be devised for the management of this function suggest the need for some further clarification of the assumptions upon which the suggestions of the preceding section rest. The total problem concept of technology assessment is the approach herein adopted. This concept involves a comprehensive information management activity having an "objective" orientation to the assessment function. It would not assume a prosecutorial, partisan stance, one primarily concerned with seeking out adverse impacts of technological applications. On the contrary, the total problem approach to technology assessment would be concerned with monitoring the performance of technology assessment systems relevant to each of the major techno-social problem areas, recommending optimum social subsystems for interim assessments, identifying opportunities for the application of technology to the solution of social problems, and seeking out and publicizing existing or prospective detrimental impacts, especially in those problem contexts where such information is unlikely to come to the attention of relevant users or other affected members of the public through normal information channels. The stance of such an activity would be neutral. Its information management functions would be comprehensive. An assumption is made here that is certainly open to debate and testing, namely, that even "shocking misapplications" of technology can more readily and reliably be identified and avoided through continuing monitoring of all significant techno-social problem areas than "hunch-sampling" techniques can achieve. In time, however, systematic selective sampling may be an adequate substitute for the total problem approach both with respect to identifying opportunities for the effective application of new technologies to social problems and identifying probable detrimental impacts of new applications.

One basic purpose of this paper has been the examination of certain notions such as fragmentation of assessment systems, the total problem assessment approach, optimum social subsystem for examination in particular assessments, and adequacy of assessments. A second basic purpose has been to pose two critical and interrelated questions: 1. What is the magnitude of the technology assessment function? and 2. What is the level of support required for the effective performance of this function? It has not been the purpose of this paper to propose a final organizational structure for the execution of the function. Hopefully, feasible alternative structures will follow from an inquiry into the two questions posed above. In this connection, the assessment centers described are primarily intended to suggest the level of effort and the quality of performance that would be required to maintain total problem assessments of all major techno-social problem areas rather than to reflect a hard and fast proposal for an organizational structure that could accomplish the overall assessment task.

It follows from the comments in the preceding paragraph that it was not the purpose of this paper to give attention to the complex of questions that arise in designing a mechanism or combination of mechanisms that might be employed in establishing the linkage between the assessment function--information management--and the official decision function--the Congressional political arena. This problem involves many intricate dimensions that are now being considered by other study groups.

NOTES

1. Wilbur H. Ferry, "Must We Rewrite the Constitution to Control Technology?," Saturday Review, LI (March 2, 1968) 50 and 54.

2. U.S. Senate Resolution 78, CLV, 19, 91st Cong., 1st Sess. Congressional Record (January 29, 1969).

3. Charles J. Singer et al., A History of Technology, IV (New York: Clarendon Press of Oxford University Press, 1958).

4. Thomas Kirston Derry and Trevor I. Williams, A Short History of Technology (New York: Oxford University Press, 1961).

5. Melvin Kranzberg and Carroll W. Pursell, Jr., eds, Technology in Western Civilization (New York: Oxford University Press, 1967).

6. Bronowski and Bruce Mazlish, The Western Intellectual Tradition (Fernhill, 1960).

7. Elting E. Morrison, Men, Machines, and Modern Times (Cambridge, Mass.: MIT Press, 1966).

8. A. Hunter Dupree, Science in the Federal Government (Cambridge, Mass: Belknap Press of Harvard University Press, 1957).

9. R. J. Forbes, in Kranzberg, op. cit., p. 29.

10. See, e.g., Alyce Edith Mange, The Near Eastern Policy of Emperor Napoleon III (Urbana, Ill: University of Illinois Press, 1940); Charles William Malberg, The Suez Canal: Its History and Diplomatic Importance (London: P. S. King and Sons, 1931).

11. Dupree, op. cit., p. 36.

12. A. Rupert Hall, "Early Modern Technology to 1600," in Kranzberg, op. cit., p. 83.

13. Eugene S. Ferguson, "Steam Transportation," ibid., p. 301.

14. Derry, op. cit., p. 548.

15. Dupree, op. cit., p. 164.

16. John G. Burke, "Bursting Boilers and the Federal Power," Technology and Culture, VII (Winter, 1966), 18.

17. Ibid., p. 13-23.

18. Dupree, op. cit., p. 154.

19. Ibid., pp. 236-38.

20. Ibid., p. 238.

21. Ibid., p. 235.

22. Hall, op. cit., p. 114.

23. Ibid., pp. 114-15.

24. Singer, op. cit., pp. 831-35. "The passing of Lord Shaftesbury's Mines Act in 1842 and of a series of Factory Acts from 1802 onwards helped to prevent some of the worst abuses."

25. Derry, op. cit., p. 709.

26. Morrison, op. cit., p. 208.

27. The doctrine of "freedom of contract" had its source in the due process clause of the Constitution. The 14th Amendment, making the due process clause of the 5th Amendment applicable to the states, prohibits the deprivation of liberty without the due process of law. The notion of "liberty" was interpreted to include "freedom of contract" in the sense (highly oversimplified) that governmental intervention would not be tolerated in the making of contracts even in those instances wherein the contract was between an individual and a huge corporate enterprise (person). Many early attempts by the states to protect industrial workers through minimum wage and maximum hour laws were defeated by the invocation of this doctrine. Innumerable articles

have been written on this topic. See, e.g., Robert G. McCloskey, "Economic Due Process and the Supreme Court," in Supreme Court Review, XXXIV, p. 34 (University of Chicago Press, P. Kurland, ed., 1962). Early "political" efforts to place some curbs on rampant industrialism included the Interstate Commerce Act of 1887, the Sherman Antitrust Act of 1890, and the active policy of Presidents Theodore Roosevelt and Taft to prosecute the "trusts." See Richard Brandon Morris, Great Presidential Decisions (Philadelphia: J. B. Lippincott Company, 1960), pp. 323-25 (including the First Annual Message of President Roosevelt of December 3, 1901). A "positive" thrust was given to due process interpretation in the case of West Coast Hotel Co. v. Parrish, 300 U.S. 379 (1937) (involving minimum wage regulation for women), wherein the Supreme Court (Chief Justice Hughes) stated (p. 391):

> "But the liberty safeguarded is liberty in a social organization which requires the protection of law against the evils which menace the health, safety, morals, and welfare of the people. Liberty under the Constitution is thus necessarily subject to the restraints of due process, and regulation which is reasonable in relation to its subject and is adopted in the interests of the community is due process."

See discussion in Arthur Miller, "Toward the Techno-Corporate State: An Essay in American Constitutionalism," Villanova Law Review, XIV, 1 (1968), 35-37 (Program of Policy Studies Reprint No. 3).

28. This theme has infinite variations, and in some respects may seem, and is, inconsistent with the prior proposition supported by active intervention of the Federal Judiciary in striking down state legislation designed to protect the health, safety, and general welfare of its citizens. Further, the Federal Government had taken control over interstate railway rates from the States and placed it in the Interstate Commerce Commission. But there have been many areas in which the Federal Government has been slow to move, such as in establishing control over adulterated foods and useless or harmful drugs. Dupree, op. cit., provides many examples. The delicate relationship that exists between Federal and state-local control over transportation activities is vividly demonstrated in statements from U.S. Senate, Committee on Public Works, Report (on S. 2658) No. 1026 (March 27, 1968), p. 1 (highway and motor carrier regulation) and the U.S. Senate, Committee on Commerce Report No. 1353 to accompany H.R. 3400 on "Aircraft Noise Abatement" (July 1, 1968), pp. 6-7 (air transportation).

29. Dupree, op. cit., pp. 177-81.

30. Ibid., p. 48.

31. For Example, the Motor Carrier Act of 1939, 49 Stat. 543; the Civil Aeronautics Act of 1938, 49 U.S.C.A. S.401; Federal Power Act of 1935, 49 Stat. 847, 16 U.S.C.A. S.824 (Federal Power Commission), (Civil Aeronautics Board); see generally, Paul G. Kauper, Constitutional Law, Cases and Materials, 2nd ed., (Boston: Little, Brown, and Company, 1960), pp. 151-55; Kenneth Culp Davis, Administrative Law (St. Paul: West Publishing Co., 1965), Chap. 1.

32. See Network Broadcasting, p. 53 (1958) U.S. House, Committee on Interstate and Foreign Commerce Report No. 1297, 85th Cong., 2nd sess. The Communications Act of 1934, 48 Stat. 1064, 47 U.S.C., S.151 et seq. (1934) was an attempt by Congress to consolidate the various statutes by which Congress had previously supervised various modes of communication. The 1934 Act, insofar as it related to broadcasting, essentially incorporated the Radio Act of 1927, 44 Stat. 1162 (1927). The Federal Communications Commission was established pursuant to the Communications Act of 1934.

33. See Louis H. Mayo, "The New Technology and National Goals: Some Implications for Legal-Policy Decision-Making," Notre Dame Lawyer, XXXVII, 33 (1961), 42.

34. The Marine Resources and Engineering Development Act of 1966, P.L. 89-454.

35. U.S. House, Committee on Science and Astronautics, Subcommittee on Science, Research and Development, Technology Assessment, Statement of Emilio Q. Daddario, Chairman (July 3, 1967), Committee Print, 90th Cong., 1st sess., p. 14.

36. U.S. Senate Resolution 78, Vol. 115, No. 19, 91st Cong., 1st sess., Congressional Record (January 29, 1969).

37. Ibid., p. 7.

38. Illustrative of this point is the following extract from the Report by the Jet Aircraft Noise Panel of OST, Alleviation of Jet Aircraft Noise Near Airports (March, 1966), p. 4:

> "In recent years an industry-wide organization, the National Aircraft Noise Abatement Council, has been engaged in developing a broad program of studies and research aimed at reduction of jet aircraft noise. However, a major

difficulty in developing practical approaches has been inhibition of initiative by any one of the several groups involved because of conflicting economic and other interests. For example, it is difficult for engine manufacturers to initiate costly engine modifications because airline operators claim they cannot afford to pay for them; for economic reasons, some airline operators insist on maximizing the payload to be carried with a specific engine-airframe combination in spite of resulting take-off noise levels objectionable to communities near the airports; some local governments in communities bordering on airports are not willing to accept the economic consequences of zoning constraints or the exercise of eminent domain and argue that aircraft noise rather be reduced by improving engines, off-loading aircraft and steeper take-offs and landings; and many affected residents object to being uprooted, and having to give up long-time personal and community associations."

39. U.S. v. Causby, 328 U.S. 256 (1946).

40. See James D. Hill, "Liability For Aircraft Noise--The Aftermath of Causby and Griggs," Miami Law Review, XIX, 1 (1964), 15.

41. Ibid., pp. 15-20.

42. Griggs v. Allegheny County 369 U.S. 84 (1962).

43. Hill, op. cit., pp. 20-2.

44. U.S. Senate, Committee on Commerce, Report No. 1353 to accompany H.R. 3400, Aircraft Noise Abatement, 90th Cong., 2nd sess. (July 1, 1968), p. 2.

45. Department of Transportation/Federal Aviation Agency Notice of Proposed Rule-Making, Docket No. 9337, Notice No. 69-1, "Noise Standards: Aircraft Type Certification" (January 3, 1969), p. 2.

46. Federal Council on Science and Technology, Committee on Environmental Quality Report, Noise--Sound Without Value (September, 1968), Sec. on "Objectives of the Task Force."

47. U.S. Senate, Committee on Aeronautical and Space Sciences Report prepared by Legislative Reference Service of the Library of

Congress, Policy Planning For Aeronautical Research and Development (May, 1966).

48. Office of Science and Technology, Jet Aircraft Noise Panel Report, Alleviation of Jet Aircraft Noise Near Airports (March, 1966).

49. National Academy of Engineering, Aeronautics and Space Engineering Board Summary Report, Civil Aviation Research and Development: An Assessment of Federal Government Involvement (August, 1968).

50. U.S. Senate, Committee on Commerce, op. cit.

51. See discussion of "technological fix" in U.S. House, Committee on Science and Astronautics Report by the National Academy of Science, Social Problems and National Socio-Technical Institutes [published in Applied Science and Technological Progress, X. Y. Weinberg, ed. (1967), p. 415].

52. Karl D. Kryter, "Sonic Booms from Supersonic Transport," Science (January 24, 1969), 359-67.

53. Jones, "Systems Approaches to Multi-Variable Socioeconomic Problems: An Appraisal" (Staff Discussion Paper, Program of Policy Studies in Science and Technology, The George Washington University, 1968), pp. 23-24.

54. Louis H. Mayo and Jones, "Legal-Policy Decision Process: Alternative Thinking and the Predictive Function," George Washington Legal Review, XXXIII, 320 (1964), 349-51.

55. U.S. Senate, Committee on Public Works Report, Vehicular Weights and Dimensions, to accompany S. 2658 (March 27, 1968), p. 1. (Hereinafter referred to as Report No. 1026).

56. Ibid., p. 12.

57. U.S. House, Document No. 354, 88th Cong., 2nd sess., Maximum Desirable Dimensions and Weights of Vehicles Operated on the Federal Aid Systems (August 19, 1964).

58. Report No. 1026, p. 12.

59. Ibid, p. 3.

60. U.S. House, Document No. 354, op. cit., p. 17:

"To solve the serious problems developed by the analysis of the available but incomplete data in the report, a policy of utilizing progressive steps to implement any findings with regard to vehicle standards has been adopted. It is contingent on the continued support by the Federal and state governments of existing levels of funding. This policy of progressive implementation provides an adequate interim solution for states, an upgrading of standards after a reasonable period on the basis of present knowledge, and a program for developing a further improved approach to this increasingly complex problem." [Emphasis added.]

61. Presidential Advisory Committee Report, A Ten-Year National Highway Program, attached to the Message of the President, "National Highway Program" (February 22, 1955), p. 1 (Referred to the U.S. House, Committee on Public Works, 84th Cong., 1st sess., House Document No. 93).

62. Ibid.

63. See U.S. House, Committee on Public Works Report 2022 to accompany H.R. 10660, on the Federal Highway and Highway Revenue Acts of 1956 (April 21, 1956) (committed to the Committee of the Whole House on the State of the Union).

64. Presidential Advisory Committee Report, op. cit., p. 3.

65. See for example, ibid., p. 24. in reference to fact that the President's Advisory Committee "made no attempt to evaluate possible revenue from rentals to concessionaires serving the traveling public nor has it attempted to estimate the additional tax revenue which will result from the creation of new values in real property resulting from the improvement."

66. P.L. 85-381, 72 Stat. 89 (1958); P.L. 86-342, 73 Stat. 611 (1959); P.L. 87-61, 75 Stat. 122 (1961); P.L. 88-157, 77 Stat. 276 (1963).

67. P.L. 86-493, 74 Stat. 162 (1960).

68. P.L. 88-206, 77 Stat. 392 (1963).

69. P.L. 89-272, 79 Stat. 992 (1965).

70. P.L. 90-148, 81 Stat. 485 (1967).

71. P.L. 85-684 (1958).

72. P.L. 87-637, 76 Stat. 437 (1962)

73. P.L. 88-201, 77 Stat. 361 (1963).

74. P.L. 89-139, 79 Stat. 578 (1965).

75. P.L. 89-563, 80 Stat. 718 (1966).

76. P.L. 89-564, 80 Stat. 731 (1961).

77. P.L. 90-495, 82 Stat. --- (1968).

78. P.L. 89-285, 79 Stat. 1028 (1965); P.L. 90-495, 82 Stat. --- (1968).

79. P.L. 89-670, 80 Stat. 931 (1966).

80. Department of Transportation, Bureau of Public Roads, _Policy and Procedure Memorandum_ (January 14, 1969).

81. The American Automobile Association launched a full page ad attack on the bill which exhibited a cartoon of a trucking "road hog" with the caption: "Mr. Congressman: Will you let the trucking lobby run over 104 million motorists?" See _Washington Evening Star_, July 15, 1968, Sec. B, back page.

82. John W. Fuller, _Current Issues in the Regulation of Motor Vehicle Sizes and Weights_, in the _Congressional Record_, CXIV, 124 (July 10, 1968), p. E. 6622.

83. Office of Science and Technology, Jet Aircraft Noise Panel, _op. cit._

84. Nor should any one agency necessarily have the _formal authority_ as contrasted with having the essential resources and the responsibility for Total Problem Assessment. _Formal authority_ here refers to the official decision function (authorization and program implementation), whereas _responsibility_ refers to the information management (assessment) phase.

DISCUSSION OF THE MANAGEMENT OF TECHNOLOGY ASSESSMENT

Dr. Mayo opened the discussion with a review of some of the more important points in his paper. The technology assessment function, he held, can be thought of as lying between two extremes. In the first, the assessment process is viewed as "highly selective, restrictive, investigatory approach," which focuses on trying to uncover the serious adverse effects or potential benefits of particular technological applications. In the other, the assessment function involves "taking a comprehensive information management approach to technology assessment which would attempt to do some initial appraising and . . . continuing monitoring of all of the major techno-social problem areas in society." Although there "clearly are any number of alternative ways in which one might go about treating technology assessment," Dr. Mayo chose to start from the board base indicated in the second extreme mentioned above.

Dr. Mayo made a brief reference to the isolation of various study efforts directed toward the understanding of the assessment process when he noted the need he had found to include some definitions of the terminology used in the paper. "It seemed to me," he commented, "that a bit of standardization of certain components of the technology assessment function is helpful in trying to analyze these problems."

He then introduced the notion of the total problem approach to technology assessment. This involves looking at a technological application with the intent of examining all of the significant impacts of that application on society. Such an approach was contrasted to the system of truncated or partial assessments, which characterize current attempts at the appraisal of technology. Truncated assessments occur because of various factors, including the limited authority of most bodies engaged in the assessment process, limited capability, and scarcity of resources. Dr. Mayo noted that

> . . . it is very infrequently . . . that we have a situation of a total problem assessment being made. We do not have entities in our technology assessment structure which have the responsibility, formal or voluntarily assumed, for total problem assessment.

The paper discussed one example of this lack of overall responsibility: the aircraft noise problem. The example of Congressional consideration of truck size and weight regulation demonstrated another failing of the assessment process, the lack of any central information centers which are capable of providing Congress with the kind of information necessary to discern a broad range of social impacts of a given technology. This in Dr. Mayo's view is a glaring deficiency but one which can hopefully be remedied: "There is a possibility that even if it is not feasible to do any substantial reorganization of the Government [and] even with the fragmentation of authority for programs . . ., the information base upon which these programs operate and upon which decisions are made . . . can be coordinated in some way." He suggested the need for new kinds of mechanisms in support of Congress to provide an effective information base. To this end he proposed a number of assessment centers with the capability to take a total problem approach to technology assessment.

THE NEED FOR A NEW ORGANIZATION

An agency representative was concerned that the paper might be attacking the wrong problem in attempting to find a way to introduce a broad, total problem approach:

> I'm not sure that the question is most fruitfully formulated in terms of whether there is any place that Congress can go to get information. I suspect that . . . if one combines the Legislative Reference Service and the resources available through National Academy of Science, National Academy of Engineering, and the National Research Council, that there is structurally no dearth of places that Congress can go to get specific answers to well-formulated questions . . . .

Congress' problem, in his view, is the lack of any "coherent focus for all of the diverse pressures that exist" so that the correct questions can be asked in a timely way. The need, then, is for an improvement in the ways that Congress formulates the questions that must be answered for effective assessment.

It was noted by a professor that it is often difficult to separate the group asking a question from the kind of answer which is received. He claimed that under the current system the "group that asks the question knows pretty well in advance what the answer will be." He felt that somehow "the group that answers the questions must be divorced from the group that asks the questions" and decides who will answer them. (A dissenter, a Congressional staff member, interjected: "Then nobody will pay any attention to the answer.")

A Legislative Reference Service Researcher agreed that the problem of framing questions is a crucial one, but he noted that one needs some basis to ask a good question and to screen frivolous and inconsequential concerns. Some information, perhaps even some research in advance of the emergence of an assessment problem, is thus necessary to allow the Congress to understand the issues involved and investigate all of the important aspects of a problem.

An industry representative, however, was not happy with the proposal to establish technology assessment centers: "When I read about a suggestion of creating a new body of any sort to add to the multiplicity of bodies we already have, I tend to throw up my hands. We have more than enough already." A Congressional Aide agreed that the concept of a new body or bodies was not workable. He saw any such body as merely another power base and claimed that Congress would never be willing to support another political power base. Another participant took issue with this view. He contended that Congressional interest and support would depend only upon "how high on the list of society's worries" technology assessment comes. In his view it is quite likely that Congress will soon recognize the magnitude of the task it must perform in this very rapidly changing world and will "someday begin to array itself with supporting institutions with quite diverse characteristics."

A Congressional Aide pointed out that the problem of securing scientific advice for Congress is not a new one. Various proposals for obtaining such advice have been brought before Congress, particularly in the late 1950's and early 1960's. The issue was resolved at that time through the creation of the Science Policy Research Division of the Legislative Reference Service. The speaker was careful to point out that such a location was chosen for the new body because of the control which Congress could have over the work done by the Division. "Congress pays the salaries," he noted, "but they [the LRS] have absolutely immutable ground rules . . . as to how they operate." Another participant didn't feel that the Division represented an adequate solution of the problem. He believed that the staff of twenty-five was not sufficient to handle the job of providing scientific information to Congress. The National Academy of Sciences (NAS) was suggested as a possible source of information; it has the advantage of being able to draw on a "broad and deep body of expertise." However, a professor questioned the effectiveness of the NAS as an assessment mechanism because "the Academy has no continuity, or very little, in given areas." An industry representative, however, found the lack of continuity "a plus, because I would hate to have the same guys passing judgment on everything."

While expressing agreement with the broad thesis of the paper, another participant commented that he "would not quite be prepared to come down strongly on the point that . . . total problem assessments don't occur under the present system." He cited flouridation and nuclear testing in the atmosphere as two examples of problems which were assessed on the issues involved. An agency representative noted, however, "that one might take the two examples mentioned and make something of a classic case study of the inadequacy of the existing assessment method." The two cases demonstrate the typical problem of assessments carried out by agencies which are committed to the advancement of given technologies, in these cases the Public Health Service and the Atomic Energy Commission. He conceded that "in neither case can one say that the system was wholly a disaster; but I think that in both cases it is fairly clear that a more objective kind of assessment might have been obtained if there were a more neutral source of funds and of evaluation." To this, the first speaker replied that he had not meant to convey the impression that technology assessment is in good shape in government, but only to indicate that he thought that, on occasion, something approaching a total impact assessment is attained. He added that he is "in favor of creating more effective and organized and focused machinery to serve society" in the appraisal of the consequences of technology.

THE ROLE OF AD HOC STUDIES

Dr. Mayo asked the group to consider the current method of commissioning ad hoc studies when assessment needs arise. He mentioned the recent establishment of commissions to study the effects of DDT and SST and asked what the purpose of such action is:

> Why hasn't somebody been in a position to have gathered [the information] . . . so that we would have the data right now on what we know and what we do not know . . . . Is there no place that you can go to find the accumulated information that gives you all the data on the prospective impacts . . . ?

Several participants did not seem as concerned with the commission method of assessment and pointed out advantages that can accrue from ad hoc assessments. An agency representative pointed out that new assessments are necessary after a change in administration because previous "alleged fact-finding studies were affected--by judgments of people who no longer hold responsibility." In disagreement with at least one other seminar attendant, he held that these reviews of the facts and conclusions of prior studies are not diversionary or stalling tactics but are the results of a new administration taking its responsibility seriously. A professor believed that repetitive assessments were necessitated by the nature of the information available. He found no fault in a system which examines "exactly the same problem over and over again in terms of new criteria and new data that are available." The "continuing process of reevaluating the same problems . . . probably is desirable." He thus took issue with the implication that if we only had a way to gather all of the information, our assessment problems would be solved. Varying criteria and new ways of looking at old information necessarily make assessment a continuing process.

One speaker, an industry representative, went so far as to suggest that the commission mechanism be the principal method of assessment: "I would propose that each question, as it comes up, should be approached by an ad hoc collection of specialists with knowledge relevant to what the problem is." In this way the problem of aircraft noise, for example, would be examined by different people from those who study the DDT problem.

PUBLIC PARTICIPATION

The discussion turned to the recurrent topic of the role of the public in the assessment process. An agency official noted that the paper did not clearly indicate to whom the assessment centers were to report, Congress or the public. It was suggested that any assessment mechanism could report to the public while at the same time its findings could be useful to Congress. The agency representative stressed the important role which the public must play. Congressional interest in assessment problems, he noted, is not without public impetus. He was thus concerned with the emphasis on Congress as the focal point of technology assessment; the public impetus for Congressional action, unless it is fostered, could conceivable die away. A brief discussion of the role of magazines such as *Environment* followed; the importance of such a public information service was acknowledged by most of those present.

THE PROBLEM OF BIAS

A Congressional Aide raised another thorny point which recurred throughout this series of seminars. He was concerned with the neutral or objective role which was proposed for the assessment centers mentioned in the prepared paper: "How would your [Dr. Mayo's] system bring out the negative effects unless the centers were, in fact, to adopt other than a weighing or neutral stance as you imply?" The resources of the proponents of most technologies enable them to marshal large numbers of facts, whereas "those who are opposed generally are in a weaker position." The speaker thought that only through the establishment of some kind of advocate for the public interest could the assessment system make certain that all aspects of a technological application would be surfaced. Dr. Mayo commented that this proposal did not preclude the establishment of such an ombudsman or advocate; that role may in fact be necessary, but it is only one step toward the implementation of a total problem approach to assessment. Another participant emphasized the need for more than one approach to the problem:

> I think the problems of technology assessment, social responsibility, and social accountability have to be exercised in a multiple context with all the limitations that go with it. You cannot lean on any one crutch . . . . [What is needed is] a variety of efforts and energies directed toward a multiplicity of critical points of power, and hopefully when the need is there, they will converge to produce an outcome that affects the quality of society.

A Legislative Reference Service Representative noted that it is crucial to recognize that technology assessment has two levels, one of which is connected with a technical problem and another which is involved in political matters. These must be separated in order to obtain effective assessments. In particular, the technical problem must be solved before the issue is thrown into the "political hopper." He claimed that this separation is really an easy one to perform if only the principals involved would put their minds to it. This point brought disagreement from several participants. Everyone present agreed with the need to separate facts from value judgments, but a university researcher noted "in order to ascertain the technical situation you have to make many assumptions which are colored by political points of view" so that the suggested separation may, in fact, be impossible. An agency official agreed; one cannot always determine the technical facts before the political process begins to operate because the answers to technical problems often come late. The separation of technical and political is impossible since often "to generate support for needed research you need political muscle."

CONCLUSION

During the discussion period an agency official sounded a warning that can serve as a conclusion to this discussion and to the series of seminars in general. He noted that "the range of possibilities in conducting a scientific investigation is almost infinite and that a scientific investigation is never conducted in a vacuum, but is always guided by some kind of conjecture." In technology assessments, the conjecture concerns the possible consequences of a new or existing technology. One must be wary of such conjecture, the speaker warned, because almost always "the problems people worry about at any given time turn out in retrospect . . . not to have been the problems they should have been worrying about." It should be the intent of those who study technology assessment to attempt to find ways of anticipating the correct and important problems which accompany the use of technology. Only then can decision-makers begin to devise policies which will maximize the beneficial impacts of technology while they minimize the detrimental consequences.

CHAPTER

6

TECHNOLOGY ASSESSMENT IN THE EXECUTIVE OFFICE OF THE PRESIDENT

Charles V. Kidd

The common problem, yours, mine, every one's,
Is--not to fancy what were fair in life
Provided it could be,--but, finding first
What may be, then finding how to make it fair
Up to our means.

Robert Browning. "Men and Women: Bishop Blougram's Apology"

INTRODUCTION

Technological change involves the Executive Branch of government in many ways. The Executive agencies generate and use technology. They influence the development of technology in industry in many powerful ways. The President shares responsibility for promoting the general welfare, and hence for the protection of the citizenry against the adverse effects of technological change. He must be involved with technology directly and indirectly, and conscious attention has to be paid to how his responsibilities in this field can best be discharged. This paper explores some of the aspects of this question. I shall use the term "Executive Office" as shorthand to cover the President and his immediate staff as well as the Executive Office as such. A distinction will be made between the immediate

*Presented December 11, 1969.

staff and the Executive Office at appropriate points. "Executive agencies" will refer to the departments and agencies comprising the great bulk of the Executive Branch of government.

Such a pervasive and important problem cannot be realistically "placed" in either the Congress on the Executive Branch. If such a misguided effort were made, it would not succeed because both branches of government have Constitutional responsibilities and all kinds of effective pressures which put them in and keep them in the business of technology assessment. Therefore, the real problems include not who has exclusive jurisdiction, but what each should do, how each should be organized to perform its function, and how the activities of the two branches should be related.

TECHNOLOGY ASSESSMENT DEFINED

For the purpose of this paper, I shall define technology assessment as the process of evaluating the effects of technological change on human beings and human institutions. This broad definition encompasses consideration of the beneficient and harmful effects of technological change. It does not assume that the overriding problems on which attention should be riveted are the harmful or potentially catastrophic effects of technological change. The definition encompasses concern with the preservation and enhancement of environmental quality, but is not confined to this important question. This definition also encompasses both of the useful categories identified in the report of the National Academy of Engineering: problem-oriented and technology-oriented assessment.

MORE EFFECTIVE ASSESSMENT AT VARIOUS LEVELS REQUIRED

As is true of many functions of government, different kinds of activities relating to technology assessment are needed at different levels of government. Each agency that uses, promotes, evaluates, controls, or is influenced by technological developments should be responsible for monitoring the full effects of the technological area with which it is concerned. This includes the assessment of and advice or action on adverse effects as part of its total responsibility.

There is in government no substitute for a conscious effort to evaluate in advance the full effects, bad and good, in all major Executive agencies of government. These agencies have the responsibility, talent, and resources to use and to control technology. Without this base, no effort at technology assessment in the Executive Office can be successful. The case for widespread responsibility for technology assessment is based on the logic that leads to our firmly fixed policy of embedding science and technology within all major operating agencies.

The Committee on Public Engineering Policy of the National Academy of Engineering takes another view of the utility of technology assessment by Executive agencies:

> In general, existing groups in the Executive Branch of the government are unlikely to be suited to the performance of technology assessments for Congress because such organizations have other primary missions, have a spectrum of skills weighted toward their specialized disciplines, may be committed to existing programs of the Executive Branch, and would have to be provided with special budgets and temporary staff to carry out the new function of technology assessment.[1]

This statement is directed towards technology assessment for Congress, but the limitations which the Committee ascribes to the Executive agencies would apparently apply to any technology assessment which they might undertake. It seems to me, to reiterate the point, that the Executive agencies are necessarily in the technology assessment business, and that efforts should be directed to making their assessments better and to assessing their assessments more critically. Even so far as technology assessment for Congress is concerned, the Executive agencies will continue to provide a large part of the technology assessment available to Congress. They should not be the sole sources.

It is true that the Executive agencies have weaknesses as far as technology assessment is concerned. They may pay insufficient attention to technology because of powerful traditions, budgetary limitations, or other reasons. The regulatory agencies, for example, have been notably deficient in both using and assessing technology. (This, incidentally, is an area within which the independence of the regulatory agencies decreases the effectiveness of any Executive effort at technology assessment. Other forms of influence and control will have to be exerted if the extensive and expanding responsibilities of the independent regulatory agencies for technology assessment are to be effectively exercised. This seems, in fact, to be an aspect of technology assessment which needs further organized exploration.)

In addition, any assessment of the effects of technological development done by any agency is likely to be both biased and limited. The bias derives largely from the constituency of the various agencies. The Department of Agriculture cannot be expected to give as much weight to the general environmental hazards generated by use of pesticides as it does to the immediate increase in costs of producing agricultural products that would result from banning their use. Agencies represent interests, and this is a fact more to be recognized than deplored.

Agencies are also limited. They have specified jobs to do, and they tend to give less weight to matters that are not their central concern. This is probably a good rather than a bad thing, since the processes and actions of government would come to a complete stop if everybody did everything.

Finally, agencies tend to be set in their ways, and this trait contributes to a widespread inability and unwillingness to generate ideas, and to foresee and prescribe things which may result in radical change--particularly changes which may reduce the power of the agency, tend to keep its budget down, force a change in the bureaucratic structure, or bring different kinds of people into positions of power in the bureaucracy.

In short, the Executive agencies have inescapable and important responsibilities for technology assessment, but their capacity for assessment is limited by many factors and must be supplemented within the Executive Branch.

COMPLETE CENTRALIZATION OF TECHNOLOGICAL ASSESSMENT INEFFECTIVE OR HARMFUL

One way that has been suggested as a means of augmenting the capacity of the executive Branch for assessment of technology is to establish a central agency which would assess all aspects of technology in a comprehensive logical manner and publicize the findings. Such an agency might also have regulatory powers. One proposal for such a group would involve the creation of an analogue to the Council of Economic Advisers. However, the case against such a comprehensive centralized institutionalization of the function seems to me to be strong at this time.

Perhaps the most fundamental consideration is that all parts of the government are affected by technology, generate technology, or both. This country has followed the wise course of encouraging every major branch of government to consider how technological advances can improve the operation of government, how technology relevant to the missions of agencies can be improved outside the government, and how problems generated by technology can be dealt with. This general policy has been, by and large, successful and it should continue to be a central guide to the manner in which science and technology are organized in Government. This is true of every aspect of the assessment of technology--how to encourage technological developments, how to set the conditions which will lead to the effective use of technology, and how to foresee and guard against the untoward effects of technological change. Complete centralization of technology assessment could be as unwise and as impractical as centralization of technology.

The losses from centralized technology assessment would include degrading the quality of the performance of agencies, a decrease in legislative support for all aspects of science and technology, and a decoupling of technology assessment from the matrix of technical economic and political realities in which it is most productively imbedded. Other untoward results would include unproductive duplication of effort, diffusion of authority, and a reduction in sense of responsibility of agencies for technology assessment.

One variation of centralized technology assessment--an agency with formal responsibility for a comprehensive examination of the future--is repugnant to me. It is, in my judgment, impossible to sustain a continuing, effective governmental organization that is completely future-oriented. The leadership would be weak because the intellectual center for such speculation is not in government. The staff would be weak because of the weak leadership. The organization would have little influence on decisions because it would not be in the streams through which decisions flow. This would tend to frustrate the staff and, over the long run, to reduce the quality of the staff and to their product. Because of attention to the future, a degree of prophecy would be involved, and the President cannot give an exclusive franchise on the future to any single group. Assessment of the future involves choices of values and goals and methods to achieve them as well as technological forecasting. Choosing values and goals is the heart of the political process, and the widest possible participation is needed. Accordingly, a governmental agency concerned solely with the future would find a substantial part of its recommendations ignored or overridden.

THE TECHNOLOGY ASSESSMENT FUNCTIONS OF THE EXECUTIVE OFFICE

Some functions of the President relating to technology assessment can be performed effectively only if his Executive Office takes an active role. In some cases, as in stating the position of the President to Congress, the role may be almost an exclusive one. In others, as in stimulating research, the function may be widely shared. Precisely how active the Executive Office will be with respect to any function depends upon so many unpredictable circumstances that the relative importance of the various functions cannot be ranked.

Initiate Proposals for Presidential Action

Proposals arising from assessments of technology will come from the Congress, from organized groups, from individuals and from agencies of government. However, it is reasonable to assume that there will be needs for governmental action that will be met only through Presidential action. This action can be of various kinds:

1. Formal or informal direction to executive agencies

2. Proposals for legislation and steps intended to assure enactment

3. Allocation of resources--money and people--through the budget

4. Allocation and reallocation of responsibilities within the Executive agencies

5. Power of appointment

6. Use of the Presidency as a platform for education and persuasion.

Such things do not just happen. If they are to be done effectively, there must be a mandate to someone or some organization in the Executive Office and enough people and money to do the jobs at hand.

Surveillance of Problems and Opportunities Involving More Than One Agency

Many technological problems and opportunities do not arise within single agencies. This is inherent in the relationship between technology and government. Science and technology develop rapidly and unpredictably. These developments often do not coincide with the functional governmental packages put together for other purposes and in response to other pressures. The fact that the impact of technology is often felt across rather than within governmental jurisdictions is inherent in the relationship and will continue. The continuing exposure of new problems and opportunities that do not fit neatly into the bureaucratic structure means that devices for coping with the problem must be considered as a continuing problem of government.

The capacity of peer organizations in the U.S. Government to resolve conflicts or to solve problems by cooperative efforts which they initiate and carry out without external influence is strictly limited and in inverse proportion to the significance of the problem. For example, the Departments of Health, Education, and Welfare and of Agriculture might cooperatively study and agree among themselves on the toxicity of DDT to various forms of life, but they could not be expected to agree on their own initiative on rules governing the use of DDT. Executive Office intervention would be required.

Not only will agencies tend to disagree on many issues involving jurisdictional issues, philosophical views, political matters such as relationships with constituents and Congressional committees, but they will at times tacitly ignore such problems or fail to attack them vigorously. This may be because no single agency feels completely responsible for a given problem or because agencies may feel that they have more to lose than they have to gain by acting aggressively in an area where other agencies may have an equal or greater claim to jurisdiction. A secretary often has more than enough problems in his own bailiwick, and does not have to search out causes which can generate difficulties with fellow cabinet members and perhaps for the President. Therefore, there exists a continuing need not only for resolving conflicts among agencies arising out of zeal to claim jurisdiction but also for identifying important problems to which the agencies are paying too little attention.

Surveillance of issues crossing departmental lines is most effective from the Executive Office.

The Long Look

Executive departments find it difficult to look ahead in any systematic manner that affects action at the top. The present tends to crowd out the future. Secretaries typically have more immediate problems than they can deal with without borrowing trouble from the future. Long-term studies are done in departments, and they are sometimes influential. But the departments cannot be depended upon as a mechanism to ensure that the major problems that seem to be developing are adequately studied and acted upon by the separate agencies. Not only is the future more remote than the present, but the long view generally results in a statement of problems and consideration of solutions that transcend agency lines. This decreases the incentives of individual agencies to look intensively and effectively into the future. Thus, the time element tends both to degrade further the capacity of individual agencies to assess the future and to emphasize the significance of a long look ahead as a central function.

Initiation of Studies

The Executive Office must depend primarily on other sources for the facts, analyses, and judgments which enter into any aspect of technology assessment carried on by the Office. The problems are so numerous and complicated that any attempt to carry out a substantial part of the evaluation by or in the Executive Office is to me unthinkable. The problem is rather one of establishing the expectation that assessments will be carried out, and of using the power of the Executive Branch to see that they are carried out. This includes the educational function of the President and his staff, establishment of Presidential task forces, mobilization of talent through such groups as the Presidential Scientific Advisory Commission and ad hoc Presidential task forces, influence over legislation, the budget, and influence or control over the policies and actions of the Executive agencies. Taken together, these can constitute a powerful arsenal.

The National Science Foundation (NSF) should have a particularly important role in supporting basic research relevant to the assessment of technology. A National Academy of Sciences Report proposes that the NSF have a Technology Assessment Division which would "by prior agreement with the Office of Science and Technology (OST), commission specifically defined assessments under contract," and administer a grant program which would identify new problems and issues and stimulate research on them.[2] The Director of the NSF testified effectively before Congress (November 18, 1969) as to the urgent need for basic research relating to technology assessment. The NSF should also conduct or commission studies of the nature and effectiveness of technologies for assessment both within and outside government. In addition, the NSF could sponsor conferences and symposia on assessment matters and prepare in-house position papers and policy recommendations. Indeed, the Director of the OST has made these recommendations.[3] This relationship between the OST and NSF indicates one of the major ways in which essential tasks relating to assessment can be done without involving the OST in their administration.

The National Academy of Engineering report said that "technology assessments should be produced in an environment free from political influence or predetermined bias."[4] This is no doubt true of the particular kind of technology assessment advocated by the Academy. This kind of objective assessment flavored by the intuitive judgments of experienced people is needed by the Executive Office. However, more than such studies are required for effective performance of the Executive Office functions relating to technology assessment. The Executive Office is an arm of the President and the Presidency, and it must be useful to the President by posing problems and solutions in terms that help him. The question is always, "What should the President do?" and not "What does a technical assessment demonstrate?" In the wise words of William Carey,

> The Presidency is weak in policy analysis. It stands perched on a bottom-heavy administrative and operational system consisting of departments and agencies equipped with resources, clienteles, and historical baggage which continually threaten to outthink and outrun the tenuous policy management capabilities of the White House.[5]

Ultimately, studies of technology assessment should be woven into the broader process of policy assessment.

While this approach is the central one in the Executive Office, there will no doubt always be specific problems that have not for one reason or another been adequately explored. In this circumstance, some part of the Executive Office may find it useful to stimulate a study. This can be done through the PSAC mode of operation, by asking an Executive agency to conduct or finance a study, or by asking some individual or group, for example an industrial group, an academic scholar, or a think tank to undertake a study. What cannot be done is to depend as a mode of operation on Executive Office personnel to conduct studies themselves.

Provision of A Receptor for Ideas

The effectiveness of technology assessment in the Executive Branch depends heavily on the free flow of facts, ideas, and judgments into government from outside. This is true at all levels of government, including the Executive Office. The problem arises not only in terms of whether there is someone to listen. There almost always is a person who will give a decent hearing to a position to propagate the facts or proposals, to get a serious hearing for them at the points where parts of decisions are made and actions are taken, or both.

The effectiveness of the listener--his influence or power, in effect--as well as his politeness and receptivity are in question. Effectiveness on such matters in the Executive Office depends upon such factors as proximity to the President, relations with the array of Presidential advisors, relations with the Executive agencies and Congressional committees, persuasiveness with nongovernmental groups and ease of access to their expertise and influence, and, least important, personal technical competence.

When a problem becomes highly important to the President, the demands on the staff are such that some kind of organized effort, as contrasted with the work of an individual, is called for. As a generalization, the more important the problem (to the President, and not abstractly) the greater the degree of formality of the organization.

Sustain Effective Links with Congress

Congress needs an improved technology assessment capability of its own, and this subject has been extensively discussed. However, Congress will always need a means of dealing with the President's Office on matters involving technology assessment. The Executive Branch will continue to carry out the laws passed by Congress, and this will continue to give the Executive Branch larger areas of discretion in dealing with technology assessment. Congress will be increasingly interested in how this discretion is used. The President will continue to take the initiative in many areas of legislation. He has extensive discretionary powers to organize the Executive Branch. His Office is the point at which much administration policy is determined, and he is always in form and often in fact ultimately responsible for the policies and actions of the departments. All of these things are obvious, but their reiteration in the context of technology assessment reinforces the point that the President must be involved and that Congress must have an effective link to the Presidency.

In all probability, the need for this link would increase in proportion to the capacity of Congress to assess technology. The broader, more fundamental and long-range the Congressional perspective, the more penetrating will be the questions directed to the Executive Branch, and the more will they be answerable only in terms of the totality of the involvement of the Executive Department and in terms of the policies of the President. The better informed and better organized Congress becomes, the greater the pressure to equip the Executive Office to deal competently with technology assessment.

In summary, as a problem becomes important on the national agenda, it is to the advantage of both the President and the Congress to have an instrument specifically designed to deal with the problem both in the Executive Office and the Congress. The President needs the capacity to formulate and present an administration position to Congress and to respond to Congressional inquiries directed specifically to him or to his office.

Capacity to deal effectively with technology assessment in the Executive Office should be viewed not as competition with Congress in this area, but rather as a prerequisite to the Presidential responsiveness to the will of the people, as expressed through Congress, and to the exercise of the Constitutional powers and responsibilities of the President

APPARATUS OF THE EXECUTIVE OFFICE FOR TECHNOLOGY ASSESSMENT

Many parts of the Executive Office are and will continue to be concerned with technology assessment. The National Security Council, the Arms Control and Disarmament Agency, the Urban Affairs Council, the Office of Emergency Planning, and the Council of Economic Advisors must assess various technologies from various points of view. This will, of course, continue. These organizations should be no more inhibited in assessing technology than they are in using technology or in promoting the use of technology. However, jointly and severally they do not constitute an apparatus competent to deal with the pressures for technology assessment which now exist.

The Bureau of the Budget has a hand in practically everything that goes on in the Executive Office, primarily because of its budgetary functions. Accordingly, it is involved with technology assessment in various ways. However, the Bureau is not the appropriate place for technology assessment functions as such. It wisely tends to keep itself in the position of asking questions rather than proposing administrative actions subject to questioning. The recent PPBS fad would encompass aspects of technology assessment, but this is not a sufficient reason to suppose that the Bureau will or should assume significant technology assessment functions. In any event, the Bureau is already overextended. Moreover, it does not have and does not want either the staff expertise or the access to expertise required for effective performance in this area.

The Office of Science and Technology has several characteristics which make it a logical place to assess some aspects of technology:

1. It has the basic staff required for the assessment of technology.

2. It has the network of relationships in the academic, governmental, and industrial world essential to identification of emerging problems, securing of advice, and help from varied sources.

3. It has had substantial experience with technology assessment.

However, the OST has weaknesses as an instrument for technology assessment:

1. It does not have a large enough staff even to identify an adequate sample of emerging problems.

2. It has no specific mandate to work on technology assessment, except in connection with the environment.

3. Many committees of Congress, including the Subcommittee of the House Committee of Appropriations, which controls the funds available to the OST, do not recognize the OST as the central point in the White House for technology assessment. (A specific, wide Presidential mandate to the OST to work on technology assessment is more important as a means of legitimatizing efforts in this area in the eyes of Congress than it is to provide the OST the necessary power to work within the Executive Branch.) The primary consequences of this are to hold the OST budget below the level required if the necessary staff is to be added, and to leave too many committees of Congress with the impression that there is no place in the Executive Office concerned with technology assessment.

4. The OST has been given so many assignments within and outside the fields of science and technology that major technology assessment responsibilities would unduly diffuse its mission.

5. Finally, and most important, the fact that the OST is oriented toward science and technology is a deficiency as well as an asset in dealing with technology assessment. Economic, social, and political factors are centrally important to many aspects of technology assessment. The process of dealing with the adverse effects of many technological changes is more a matter of dealing with economic, political and sociological factors than of finding a technological remedy. The OST is widely considered as overstressing the technological aspects

of technological assessment, of being overly influenced by the scientific and technological establishment, and of lacking the staff skills required for a comprehensive approach to technological assessment.

In summary, the Executive Office is not well enough equipped to deal with the assessment of technology. By "not well enough" I mean that there is a clear and present danger that a number of responsibilities of the President in this area will be poorly handled because his Office is inadequately staffed and organized.

ENVIRONMENTAL QUALITY AND TECHNOLOGY ASSESSMENT

The degradation of the quality of the environment, largely a consequence of undesirable and unanticipated effects of technological "advances," is clearly the area of technology assessment most significant on the national scene generally and in the Executive Office in particular. Other consequences of technological change may later become of equal or greater significance, but the environment will be the most important area of technological assessment for some time to come. Accordingly, the adjustments to deal with this problem, which can be only briefly sketched, are significant.

The rapidity with which the structural changes noted below followed the emergence of acute popular and Congressional concern over air and water pollution, and other forms of deterioration of the environment is significant. These changes lend further credence to the old homily that changes generally follow rather than precede the emergence of a crisis.

The Environmental Quality Council

In connection with the evolution of the Executive Office structure for technology assessment, the establishment of the Environmental Quality Council in May, 1969, was an important event. The salient facts are these:

1. The Council was established by Executive Order, thus giving it direct Presidential sanction.[6]

2. The Council is a Cabinet level group over which the President presides. (The members are the Vice President and the Secretaries of Agriculture; Commerce; Health, Education, and Welfare; Housing

and Urban Development; Interior; and Transportation. The Director of the Bureau of the Budget, the Chairman of the Council of Economic Advisers, and the Executive Secretary of the Council for Urban Affairs are observers).

3. The Science Adviser to the President is the Executive Secretary of the Council and assists the President in directing the Council. The OST provides staff and administrative support for the Council.

4. The Council has broadly defined responsibilities:

> [It] shall advise and assist the President with respect to environmental quality matters and shall perform such other related duties as the President may from time to time prescribe. In addition thereto, the Council is directed to:
>
> (a) Recommend measures to ensure that Federal policies and programs, including those for development and conservation of natural resources, take adequate account of environmental effects.
>
> (b) Review the adequacy of existing systems for monitoring and predicting environmental changes so as to achieve effective coverage and efficient use of facilities and other resources.
>
> (c) Foster cooperation between the Federal Government, state and local governments, and private organizations in environmental programs.
>
> (d) Seek advancement of scientific knowledge of changes in the environment and encourage the development of technology to prevent or minimize adverse effects that endanger man's health and well-being.
>
> (e) Stimulate public and private participation in programs and activities to protect against pollution of the nation's air, water, and land and its living resources.
>
> (f) Encourage timely public disclosure by all levels of government and by private parties of plans that would affect the quality of environment.
>
> (g) Assure assessment of new and changing technologies for their potential effects on the environment.
>
> (h) Facilitate coordination among departments and agencies of the Federal Government in protecting and improving the environment.

In addition, the Council is responsible for the plans and actions of Federal agencies affecting outdoor recreation and natural beauty.

Finally, the Council is directed to assist the President in preparing periodic reports to the Congress on the matters covered by the Order.

These powers and responsibilities are quoted at length because they constituted the first specific, broad, concrete, continuing recognition of the significance of technology assessment in the Executive Office. Without unduly stretching the concept of the environment, certainly most of the areas of technology assessment that have been identified as most urgent could come under the purview of this Council.

However, there are other views of the motives for establishing the Council and of its effectiveness. In Congress there was a widespread impression that the Council was established to create the impression without the substance of Executive Branch action, to keep initiative in the hands of the President and away from Congress, and to emphasize coordination rather than action and the resulting large expenditure. Certainly there was protracted jockeying for political position by the White House and Congress--notably, Senators Jackson (D., Washington) and Muskie (D., Maine). Both of them introduced bills dealing with the protection of the quality of the environment.

The Statutory Council on Environmental Quality*

On January 1, 1970, President Nixon signed the National Environmental Policy Act which created a three member Council of Environmental Advisers. This law had its origins in Congress, and particularly from Senator Jackson and Congressman Dingell (D., Michigan). Enactment of the measure over opposition from his Office and from several Cabinet members, based on the view that the Environmental Quality Council was an adequate arrangement, bears witness to the rapidly rising political potency of the issue of deterioration of the environment and to the tendency of Congress to take the initiative on the issue. In terms of the President's Office, the legislation indicates that he does not in fact have a completely free hand to organize his own staff, and that the structure of his Executive Office has important symbolic political significance. Just what the relationship between the President's Cabinet Committee and the new statutory Council will be--and whether the Cabinet Committee will survive--remains to be seen.

---

*This section was added after the seminar.

The complicated minuet has not yet been played out. The President, on signing the bill, said:

> In the near future I will forward to the Senate names of highly qualified individuals to help both the Cabinet and me in the critical decisions that will affect the quality of life in the United States for years to come. I will then take the necessary executive action to reconstitute the Cabinet Committee and its staff to avoid duplication of function.
>
> On the latter point, I know that the Congress has before it a proposal to establish yet another staff organization to deal with environmental problems in the Executive Office of the President. I believe this would be a mistake.
>
> No matter how pressing the problem, to overorganize, to overstaff, or to compound the levels of review and advice seldom brings earlier or better results.

The last sentence of the statement was directed at Senator Muskie's proposal, embodied in the Water Quality Improvement Act of 1969, which has been passed by both Houses of Congress, to set up an Office of Environmental Quality in the Executive Office. The sentiment of Congress on this specific issue, as contrasted with the traditional Congressional reluctance to provide adequately and directly for Presidential staff, is to ensure that the President's Office is amply staffed and compactly staffed. Moreover, it is reasonable to assume that Senator Muskie has a stake in seeing his name attached to a law designed to protect the quality of the environment.

THE FUTURE

It now seems clear that restoration and protection of the quality of the environment will be a major national priority for the decade of the 1970's and that numerous important political issues will center around this problem. Just how broadly "the environment" will be interpreted by the new Council remains to be seen, but the probability is that the term will be widely rather than narrowly construed. If so, most of the technology assessment functions of the Executive Office will be carried on by the Council and its staff.

The precedent set by the establishment of the Council of Environmental Advisers suggests that in the Executive Office over the long run it may be more effective to set up a separate organization and also to define the task in terms of what is to be protected--the environment, the consumer, etc.--than in terms of technology. There are plausible reasons for thinking that this approach may be preferable:

1. It is useful politically to have a symbol of action against a concrete danger. This is provided better by defining at least some parts of the task in terms such as "protection of the environment."

2. Social, economic, and political factors influence the development of hazards and they can promote or frustrate efforts to mitigate hazards. It may be more effective to organize in terms of problems to be attacked--threats to the environment, for example--than in terms of the technology involved. The former approach can facilitate an attack that encompasses but is not limited to technological factors.

3. The range of tasks assigned to the Office of Science and Technology is so broad that assumption of responsibility for a set of problems requiring expertise in many areas not closely linked to technology--the legal and administrative framework within which the Federal, state, and local governments approach problems of environmental protection and balancing the legitimate and incompatible claims of various interest groups--may unduly diffuse the efforts of the OST and do less than full justice to the problems of the environment.

4. Organization in terms of problems preserves flexibility. Organizational adjustments can be major or minor, depending on the problem, and the adjustments need not be irrevocable.

5. A problem approach is "convergent." That is, existing knowledge and the contributions of existing organizations can be brought to bear sharply on the solution of a defined problem. This focusing of effort can be done well by the Executive Office. In contrast, an approach which stresses the assessment of technology-initiated problems tends to be divergent and open-ended.

The establishment of the new Council will certainly postpone for a long time if not forever the proposal that a major unit devoted to technology assessment as such be set up in the Executive Office of the President.

The fact that one cannot be clear and unequivocal in foretelling the structure of the Executive Office reflects its nature. It is designed to help the President; and the changing, unpredictable pattern of forces and pressures that assail the President are for the most part the same forces that make it impossible to be dogmatic about the structure of his office. Only one thing is clear to me. That is, any effective organization must be in the stream of power. If functions relating to technology assessment in the Executive Office are not linked to the power system in both the Executive and Legislative branches, they will be ineffectively performed. The new Council is linked to that system.

NOTES

1. National Academy of Engineering, Committee on Public Engineering Policy Report, A Study of Technology Assessment (July, 1969).

2. National Academy of Sciences Report, Technology: Processes of Assessment and Choice (July, 1969), p. 97.

3. U.S. House, Committee on Sciences and Astronautics, Sub-Committee on Science, Research, and Development, Statement by Lee DuBridge (December 11, 1969).

4. National Academy of Engineering, op. cit., p. 3.

5. William D. Carey, "Presidential Staffing in the Sixties and Seventies," Public Administration Review, XIX, 5 (September-October, 1969).

6. Executive Order No. 11472. (May 29, 1969).

DISCUSSION OF TECHNOLOGY ASSESSMENT IN THE EXECUTIVE OFFICE OF THE PRESIDENT

Dr. Kidd began the session by explaining that he was operating, in his paper, from the assumption that the President *must* be involved in technology assessment. Thus the question which should be addressed is "not whether, but how and through what mechanisms" the President can make his involvement most effective.

Currently, most Executive Branch assessments are performed by the various agencies. But,

> It is obvious . . . that the individual agencies of government are not capable of doing the kind of job that needs to be done in technology assessment . . . because an executive agency is, by its nature, biased and partial in the examination that it gives to problems and, by and large, more inclined . . . to look backward than to take a long look ahead.

Dr. Kidd insisted, however, that he was not advocating the centralization of all assessment activities. In fact, he urged the strengthening of assessment capability within the Federal agencies. But he did feel that an extension of the role of the Executive Office in this area might alleviate some of the problems of agency assessments. Further, he claimed, some elements of the assessment process can only be performed within the Executive Office. These include: "the initiation of proposals for action by the President," the "surveillance of agency action," the coordination of agency action," the assurance that "a long look in time will be taken" since "the longer the look ahead, the less are the problems . . . likely to fall within any one agency," and the maintenance of links to Congress which will allow the Congress to know what the President thinks about particular issues.

Of course, the Executive Office, as presently structured, has some weaknesses which severely hamper its ability to perform adequate technology assessments. Two of these stand out as critical. First, the Executive Office "doesn't have any formal charter to deal with technology assessment, and that tends to make the undertaking more of less of an ad hoc undertaking." Second, it was pointed out by Dr. Kidd, the President's science apparatus doesn't have enough people to perform the tasks that ought to be performed.

Dr. Kidd felt strongly that technology assessment within the Executive Branch cannot be improved through a major reorganization of the agency structure of the Government. Any new organizational structure would fail because the process of technology is unpredictable and problems arising from the application of technology invariably cross bureaucratic lines: "regardless of what the organization is, you will have advances in new and unpredictable directions which will always force a consideration within the Executive Office of tasks which cross departmental lines."

Dr. Kidd noted that the new Environmental Quality Council represented a movement toward more effective assessment within the Executive Office:

> I consider that the establishment of the Environmental Quality Council by executive order . . . [with] a clear delineation . . . of a set of specific responsibilities having to do with assessment of problems relating to the environment . . . is a very significant event in the evolution of the structure and function of the Executive Office for technology assessment.

Although Cabinet committees are not generally known for their power or durability, they may become effective when they have a clear mandate for action.

However, with the exception of the new Council, the Executive Office is deficient in its technology assessment capability; in fact it is "deficient to the point where the President . . . does not have the apparatus, the information, or the ability to stay abreast, let alone ahead, of problems that may hit him suddenly."

Although Dr. Kidd felt that the Environmental Quality Council was a step toward improving technology assessment in the Executive Office, he was not enthusiastic about the potential advantages of new organizational structures within that Office:

> I think caution is indicated in proliferating organizations within the Executive Office. It seems to me that the capacity of the Office of Science and Technology to contain this function and to deal with it adequately has not been tested.

THE ROLE OF THE EXECUTIVE OFFICE

A Government official agreed that technology assessment must have a place within the Executive Office since "in order for this [technology assessment] to have any utility, it has to be attached, in some way or another, to the sources of power that make the decision." Any assessment mechanism within the Executive Branch which is not close enough to the President will fail to make meaningful inputs into the Presidential decision-making process. "There is a need for the Executive Office to operate some kind of an assessment mechanism which is separate from and independent of that in the rest of the executive system." This, of course, does not negate the need for assessments by the various Executive agencies and departments, but it recognizes that each of these agencies and departments has its own peculiar biases or constituencies as well as limited access to the President.

Nevertheless, some of the participants were a good deal more skeptical of the ability of the Executive Office to perform an adequate job of technology assessment. An industry representative felt that any setup within the Executive Office structure would be too susceptible to pressure from the President. He brought up the case of the assessment of the SST; "very eminent specialists from the National Academies" assisted a Cabinet level committee in "assessing this machine and found it wanting." This result was not acceptable to the President, so "when the crunch came . . . the pressures were such that the report [of the committee] simply got squelched." He generalized from this example to the conclusion that "any time you run into a controversial public policy issue . . . and you don't come up with the answer the President wants, it's too bad." Another participant, a former Government official, felt that this problem was most relevant in connection with ad hoc groups within the Executive Office:

> When you ad hoc anything within the Executive Office you are quite likely to be engaged in a very risky enterprise, because whatever the pathology of Cabinet committees might be (and it is acute) it is nothing compared to the pathology of ad hoc groups set up to attack particular problems. There the Presidential foot typically comes down . . . and just squashes them like bugs

if the President is dissatisfied. It is the ad hoc process that is unsatisfactory. A more permanent function within the Executive Office might be more effective. But even a permanent assessment body might fail. This participant characterized the job of the Executive Office as "erratic; impulsive; and the product of elbowing, jostling, opportunism, and crises." . . . [This] reminds me of that electric news sign in Times Square. The words follow each other and chase each other around and around and pretty soon they disappear somewhere on the side of the building. And that's the way it is in the Executive Office. Problems suddenly come up; you see them coming around the corner and you watch them, and all of a sudden you realize that they're being chased by other problems. And the first problems just fall off the end; it's always going to be that way in the Executive Office.

What is needed for technology assessment is "a fixed locus of concern on a continuing, systematic basis" and this may require either the altering of existing institutions or the emergence of new ones.

A university professor was certain that technology assessment within the Executive Office, and in fact within the whole Executive Branch, could not be an effective undertaking:

> I think it's totally unrealistic to expect or believe or contemplate that any technology assessment by the Executive Branch of the Government will meet the kinds of standards and perform the kinds of functions that we expect of technology assessment . . . because the Executive Branch . . .

> simply is not able to perform the function of technology assessment in the open. . . . Secondly, it is simply not capable of performing the kind of technology assessment which is likely to involve political controversy. I would say that whatever the Executive Branch does in going through the motions of technology assessment is all to the good, but we just want to reconcile ourselves to the fact that we just can't expect very much of it.

Two participants contended that Congress was the only place within the Government where technology assessments could be expected to produce meaningful results. An industry representative felt this way because Congress is "closer to everybody than is the Executive" and a university professor concurred: "I suppose that it is fitting that assessment be performed in Congress because technology assessment is really a kind of legislative or political policy-making function." He added that "if Congressman Daddario or some other member of the Congress had never come up with the question of technology assessment, the odds are very, very good that nobody in the Executive Branch ever would have come up with it."

A university researcher noted that there was really no choice to be made between assessments in the Executive and in the Congress since assessments must be made in all segments of the Government as well as in the private sector. He then commented upon some of the elements of technology assessment which Dr. Kidd had claimed could only be performed within the Executive Office. The initiation of assessments can, in fact, be performed by the Executive Office, but "anybody can do this in one form or another. I don't think that's a unique characteristic." The Executive Office is capable of taking a long look in time, but then so is the Library of Congress or a university. On the other hand, the function of pulling together the overall Executive Branch viewpoint for the information of Congress and the public is, in his view, a function which can be uniquely performed by the Executive Office. This discussant mentioned an additional element of the technology assessment process which falls within the province of the Executive Office and which could tend to alleviate one of the problems resulting from assessments within agencies or departments:

> I think that on complex technology assessment issues you are bound to have conflicts among the assessments performed by individual agencies of the Government because of their different constituencies and biases. Now it is in

> the resolution of these conflicts that the Executive Office of the President has a very unique function.

However, this speaker felt that the function of conflict resolution could only be performed with an adequate staff and that currently there is no entity within the Executive Office which is adequate.

THE ROLE OF THE OFFICE OF SCIENCE AND TECHNOLOGY

The potential role of the Office of Science and Technology in technology assessment was discussed at some length by the seminar participants. A government official felt that the effectiveness of OST was highly dependent upon the relationship between the Director of the Office (the President's Science Adviser) and the President. The role of the Director "varies from one administration to the next. There are times when he, as an individual, is an intimate of the President and there are other times when he's very much at arm's length and participates in events that go on only on formal occasions." A problem arises because of the varied roles he takes on; as the Science Adviser he is an advocate of science while as Director of the OST he is expected to be more impartial. Thus, in a sense, "he is a judge, but he basically tends to be, and is looked upon, and is wanted to be, an advocate" for a certain point of view. In addition, the OST tends to examine crisis situations in a short time frame, so that "the best you can hope to do . . . is to see that the largest fraction of existing knowledge is brought to bear in the assessment process" and there is little reason to expect that significant new information will be discovered.

A university researcher expressed some reservations about the OST as an assessor:

> I think the thing that bothers me about the OST handling the bulk of the assessments in the Executive Office is in fact that the OST does have such close ties to the scientific and technical community. . . . I think that the technology assessment problem is such that it is important that it not be approached from . . . an elistist point of view and too narrowly from a science and technology point of view. . . . Apart from the fact of its the OST's] not having adequate staff, I think the OST is much too narrowly aligned.

Another participant expressed a similar point of view. He felt that the OST

> . . . will rely on people who are very heavily technology-oriented. That, to me, is an unfortunate bias for doing technology assessment because it. . .takes the point of view that the technology is the important aspect . . .

rather than emphasizing the impact of technology on society. This technological bias is also to be found in the National Academies of Science and Engineering. "What we should be looking at is: what is the response of society; what will happen to society?" Because the OST is not capable of taking such an approach, this participant felt that "the OST, as an example of an Executive Branch agency, is not a suitable one for the conduct of technology assessment."

Still another speaker expressed some doubts about the effectiveness of the OST as an assessor because of political considerations. He claimed that

> the perceptions of the OST in Congress are that it cannot do the job. . . . It cannot do the job because Congress won't let it do the job--. . . . Congress has never been willing to go along with an expansion of the OST.

Similar dissatisfaction with the OST exists, he said, within the Bureau of the Budget, the AEC, NASA, and the Department of Defense. The widespread dissatisfaction led him to the conclusion that the OST should not be a lead organization for technology assessment. On the other hand, this speaker felt that the OST's basic problem stemmed from the lack of any mandate to act as an assessor and that the problem might be alleviated by "some kind of a designation, a mandamus, on the part of both the Congress and the President."

One participant felt that the National Goals Research Staff (NGRS) might be more suitable for technology assessment than other parts of the Executive Office because of its sensitivity to the needs of society. While he didn't feel that there was any perfect apparatus now in existence for technology assessment, he thought that the NGRS was "the closest thing I can think of right now." He wryly added that the NGRS is "enough in the primitive stage that we can call it good so far." But a university researcher felt that the NGRS could not really perform technology assessments. This was so because of limitations in manpower and because

> . . . a group that is dedicated to attempting to delineate alternative goals and to looking at the implications of various goals . . . shouldn't be mixed with a group that is trying to evaluate actual developments. I . . . think the two are conflicting functions. . . .

OTHER CONSIDERATIONS

As always in these seminars, a portion of the discussion was devoted to general considerations of technology assessment which were broadly related to the session's topic.

One speaker was somewhat concerned with the fact that some people seem to view technology assessment as a means for substituting a new elite, the scientists, for that body of experts which currently makes policy. He felt that "the science community is really saying that you really ought to submit decisions to our kind of judgment while, in fact, "by virtue of anything I've ever seen and by virtue of everything you've ever seen, the ultimate judgment is a political one." But another speaker had a nearly opposite concern. He felt that technology assessment often showed a bias against science and technology. He decried the "widespread bias against planning, which is unfortunately imbedded in the American tradition." Since many decisions which have an important bearing on the interaction of technology and society take place in industry, it is important to avoid a bias in favor of industry. But we should not go too far in the other direction and allow technology assessment to take an anti-technological course.

A Government official returned to a matter which had surfaced numerous times in the course of these seminars. He noted that the expertise necessary to understand the technical aspects of the policy formation process "tends to be concentrated among the technology advocates". He continued:

> Somehow or other it is this expertise which we would like to convert into a somewhat more impartial thing called assessment. There is . . . a wish that the expertise should be equally available to the advocates and the skeptics and that it should be public in the sense of being equally available. And that seems to me to be the real prodecural question because . . . it is very strongly built into human motives that advocacy is the thing that leads to expertise. People get to be experts because they have this very strong feeling of advocacy. . . . We should . . . search for some

> new kind of process . . . that would guarantee the equal accessibility of expertise.

On another point a participant felt that major efforts of "monitoring and surveillance" and other attempts to build an information base might not be as important as they seem. In many instances, the problem is clear-cut and what is called for is action rather than study. Technology assessment, if it is to be useful, must be the antecedant to effective action.

CHAPTER

7

THE SOCIAL FUNCTION OF TECHNOLOGY ASSESSMENT

Franklin P. Huddle

INTRODUCTION

The year is 1970. There is something about the decennial that encourages perspective thought. Although purists may define it as the closure of a decade, the popular view is of a fresh beginning. This particular year, 1970, has special significance. It is still deep in the present century, a century that has witnessed more radical changes in the condition of man than all our previous centuries together; at the same time, 1970 is within striking distance of the year 2000, the close of the second millenium of the Christian Era. Will we live to see it? Will we want to?

The dangers and the opportunities that sum up the changes in the condition of man between 1900 and 1970 derive from the exploitation of a single assumption: that of the relationship between cause and effect; in short, the scientific method.

Bit by bit, the scope of application of the scientific method has eroded the territories of the normative world. Copernicus and Galileo took away the theologically oriented Ptolemaic system and replaced it with the scientifically oriented solar system, with man no longer the center of the universe. Newton and Einstein converted the mysterious universe into an orderly workshop, with firm and inflexible rules of motion and relationships of space, time, and matter. The sun, as object of ancient worship, became a nuclear factory of

---

*Presented January 22, 1970.

electromagnetic energy. Pasteur and Koch converted disease from a visitation of God to species of controllable microorganisms. Franklin removed the mystery from the shafts of Jupiter and the hammer of Thor. The stream that gushed from the rock at the touch of Aaron's rod, the dew-pearled morning, and the thunderheads over Mauna Loa are all accounted for in terms of the hydrologic cycle.

Now, in our own time, Watson and Crick have explained the mysteries of inheritance, of blood line, of racial difference, in the understandable terms of a complex but orderly organic molecule. The heart, enshrined in our language as the seat of all emotion and loyalties, was converted by Harvey into a clever pump, and by Christian Bernard into a replaceable component.

The emotions themselves are being systematically turned into electrochemical phenomena. Instead of Coleridge's stately pleasure dome where Alph the sacred river ran, the scientist shocks a designated area of the cerebral cortex and a beatific grin spreads over the face of the chimp. Up pills and down pills are a comic-strip commonplace. As we unravel the complexities of the electrochemical linkage of man's command-and-control system, we begin to be ambivalent about the ancient virtues. What is it to be "a man" when the surgeon can convert man into woman or vice versa without any notable effort? What is it to have "courage" when a drug can give it or take it away? What is it to have "nerves of steel" when these are obtainable at the pharmacy?

The first organic molecule was synthesized in the nineteenth century; in 1969, life itself, the living molecule, was assembled by human manipulators in the laboratory. Yet this laboratory product is held so sacred in the Hindu religion that its devotees are unwilling to restrict its numbers to save their very lives.

The Christian ethic, built on the Judeo-Christian tradition of the Garden of Eden, original sin, and the curse of God, provided alike the Catholic devotion to the salvation of the soul through earthly penance, and the Protestant devotion to the salvation and rise of the individual through purifying toil. In the legend of the Garden of Eden, man by acquiring knowledge was condemned by God to toil: "In the sweat of thy brow shalt thou eat bread." Today, by virtue of his accumulated knowledge, man is freeing himself from much of his traditional serfdom to toil. A bale of cotton once took 200 hours of hard manual labor to produce; today it takes less than 14 hours of relatively light mechanical skill. We have invented machines to do our work and machines to make the machines with; we are training computers to control the machines, to make engineering decisions in the operation of the machines, to make design decision in the construction of more sophisticated new machines, and to make management decisions as to the allocation of resources in the corporations that own the machines. We are on the threshold, technologically speaking, of a world in which the goods we consume are no longer related in quantity to the hours we toil. Man may choose to work, but it is no longer the price of his survival.

We may as well begin to face up to the fact that the steam has gone out of the old debate over hours of work. Business economists and labor economists will still wrangle for a long time into the future over the relationship between productivity and wages, but the issue is a dying one. Despite all our clever excuses for reducing the workforce by eliminating child labor, extending the years of schooling, accelerating retirement, expanding military forces, sending as many young people as we can abroad voluntarily or otherwise, multiplying holidays, and sneaking in on-the-job training courses, plus the great boon of Parkinson's Law, our machines keep on producing more than we can consume and our computers continue to invade territory that was once the exclusive province of the human intellect.

Notice the ambivalence: We need more of us as consumers but fewer as producers in our technological world, while in India and Egypt the conditions clamor for more producers and fewer consumers. Yet in both worlds we are destroying the environment that alone renders existence tolerable. And is the achievement of a balance between production and consumption the glorious objective toward which all mankind should strive?

The prime fact of the twentieth century is that our social and religious rules and beliefs of metaphysics and ethics are dramatically undercut by our science. At one stroke we have wiped out the basis of our belief in economic justice. And our principal competitor, Marxism, is merely a primitive alternative to primitive capitalism; both systems have undergone modification and will require much further modification to be compatible with the changes in total human environment resulting from man's science and technology.

It is absurd to debate whether man was better off or not in the Garden of Eden, figuratively speaking, or in our modern House of Science. The rules man makes for his own behavior, and the attitudes man develops, are shaped by his environment. Man was no more happy in 1969 because the atom had been split into 200 different particles and leaded gasoline was removing the knock from high compression engines, than he was in the old stone age when rocks were selected rather than shaped as tools. Man, the primitive tool-user of 100,000 B.C., could no more foresee what was in store for him than we, today, are able to estimate how joyous we can expect to be a century hence. Each age has its own rules, mores, values, attitudes, expectations, joys and sorrows, and contributions to change. Our standards are relative and not absolute. Man's attitude toward his compatibility with his environment and his attitude toward the effectiveness of his social arrangements to preserve and improve this compatibility in the light of environmental change determine the extent of human satisfaction. Our own age of science, because it has altered our environment more rapidly and more profoundly than ever before in human experience, has made us more self-conscious than in past ages of how outmoded our ideas and values have become. We could plausibly argue, in short, that human satisfaction in our own time is at a low ebb precisely because environmental change has proceeded so rapidly that human institutions have been unable to keep pace.

Of course, our life is not all that black. We derive some extent of happiness from the exhilarating experience of change itself. We enjoy the fresh novelty of a colored television receiver, the rush of a supersonic transport, the responsiveness of an auto with higher torque and power steering. By our actions we demonstrate that we are willing to respond to the advertisements, by paying the economic costs of these goods.

The gratifications of our desires for novelty and convenience are modifying our environment in ways we want. But they are also taken together, modifying our environment in ways we do not want. Moreover, the increasingly technical nature of our culture is imposing on us the need for more and more rules, heightened standards, restrictions, and constraints.

Our proprietary drugs, for example, are so much more potent and subtle in their effects that it is not sufficient to measure their statistical safety. Today, we need to know as much as possible about their side effects and their effectiveness against specific disorders, so that we can decide the trade-off, determine whether on balance they are useful enough for us to tolerate their inevitable faults, and hidden risks. Moreover, we cannot tolerate in a headache remedy the degree of risk permitted in a drug effective against uremia or cancer.

In its effects on the environment, our technology is aggregative. The carbon dioxide from our vehicles joins the carbon dioxide from our smokestacks. The sulfur dioxide from our power plants joins the sulfur dioxide from our refineries. We learn not to wash the aspirin tablet down with Coca Cola. We mustn't drink cocktails along with our antihistimine tablets. We must choose carefully the liquid we put in our windshield spray tanks and the detergent we use in our washing machines. The rules man has to learn to adapt his behavior to this brave new world are complex and multiplying.

Superimposed on these aggregative impacts, plus and minus, of technology on the human environment is the fact of population increase. Technology--of public health, of agricultural production, of living standards, of medical care, of environmental control--has brought an unprecedented growth in numbers of people on the satellite Earth. This increase, in its turn, has generated expanding markets for the products of technology: durable consumer goods to convert material resources into litter, transport hardware to pour forth effluents to pollute our air, household supplies that add to the wastes of glass, paper, organic phosphates, and toxic chemicals whose disposals further pollute our air, water, and land.

Numbers of people living together in congested cities require more detailed rules of behavior for their safety and convenience. Persons using modern technologies also find necessary new and elaborate codes of behavior. Only thus can we drive more or less safely in our superhighways, operate electronic communications over our crowded electromagnetic spectrum, set standards for percolation of outflow from suburban septic tanks, determine suitable noise levels around crowded airports, or cross a street in safety.

"She walks in beauty, like the night." But not in Washington, D.C., where newspaper box scores of nightly violence record robberies, burglaries, rapes and aggravated assaults in a crescendo that testifies to man's inability to design an urban culture of quiet, peace, or beauty. Technologists are designing better safes and intrusion sensors, while other technologists are designing ways of defeating them.

Is it a myth that we once knew how to live together, or has man always been unfriendly to man? As the ancient Greek civilization began to collapse in Athens, the philosopher Plato called for a new kind of organized society as an antidote to violence. In a similar environment of unrest, the philosopher Hobbes declared that logic called for a similar totalitarian regime. Both were responding to the apparent needs of their environment.

It is asking a good deal of man to expect him to find his way through the tangled thicket of fact and myth in our year 1970. How much fact and how much myth will our organizations of government compel us to accept and live by? How much mendacity will we tolerate, or require ourselves to accept from Madison Avenue? When does "freedom of speech" remain a human right and when does it become a degradation of the "signal-to-noise ratio"?

One can find in his heart great sympathy for those of our young people who are "copping out" in the face of all these confusions. What, after all, is the significance of facial hair? Why is it that the Saints of our hagiology who saw visions in the desert are revered while the takers of hallucinatory drugs merit only condemnation?

It appears that science has invalidated all our traditional social guidelines and left us none in their place.

Margaret Mead, Kenneth Boulding, and many others have called our attention to the fact, so beautifully dramatized in the recent photograph from Apollo 12, that mankind are all passengers on Spaceship Earth. Seen at a distance, it is still a beautiful blue craft. The desecrations wrought by man's technology are not apparent at a distance of 100,000 miles. Even on Earth itself, the despoliation mars only a small fraction of the whole. Our fears are for the future, as we project statistically the disasters that we can foresee if we keep on doing what we have been doing, with more and more of us doing more and more of it.

There is a whimsical remark, popular in the aerospace industry, that the only thing wrong with man's Spaceship Earth is that it was delivered without a manual of operation. Here, in a way, is the nub of the matter. We _do_ need a manual of operation for our spaceship. But we also need a code of behavior for the passengers. So far, our ecological sciences have been unable to provide the one, and our social sciences have been unable to provide the other. Is this inability inherent in science itself, or are the obstacles those remaining from our prescientific past?

What are the obstacles to man's achievement of compatibility with his environment? In one sense, it is that technology has given us too much too soon. It is not easy for man to change quickly the rules he lives by. It is painful for man to relinquish the binding energy of his society. We cannot keep our sanity and change very fundamental social rules and beliefs and religions casually, from day to day, to keep them consistent with the latest findings and even hypotheses of science. We cannot by ukase command our scientists and technologists to observe periodic moratoria on discovery and imagineering.

In recent years we tried the approach of setting ourselves a difficult national goal and hiring an assembled team of expert specialists to carry it out for us while we watched their progress on our colored television sets. We pieced out the time by watching Joe Namath and several hundred counterparts drop back to pass, in the teeth of onrushing linemen. We made spectator sports technical enough to satisfy many levels of intellect. Other spectator sports like political conventions, the Vietnamese War, the gruesome array of television fare, riots, assassinations, and the war on crime are our diversions. But they scarcely serve as goals for our society. Nor are our goals as individuals challenging. Before us are such unattractive models as the recluse billionaire, the Olympic champion turned entertainer or prizefighter, the disillusioned ex-President, the charismatic politician turned poet, the space explorer selling American railroads. Most of the Horatio Alger idols of the past are today objects of ridicule. What is it good to aspire towards, for us as individuals and for us as a nation and a world?

We need new goals, but as a society and as individuals. We need as a society to develop rules of behavior to maintain and improve our compatibility with our environment. The rules and the goals must be compatible with each other, not only in the present but for the future, under the stresses of a rapidly developing technology. What are the principles that must govern our approach to this set of requirements?

If our goal-oriented society is to better its condition as a society, goals need to be adopted and pursued. But setting goals is far from simple. Among the criteria that appear to be necessary are, at a minimum, the following:

--The goal proposed must command general respect.

--It must be generally regarded as worth doing.

--It must be difficult but not impossible.

--It must not be in overt conflict with potent residual myths.

--It must be arduous but not exorbitantly expensive.

--It must offer opportunity for wide participation.

--It must offer a durable general motivation.

--Progress toward it must be measurable in understandable terms.

--Its outcome must be judged desirable.

--And, in practice, the outcome must be tolerable.

This last-named criterion is a profoundly complex one. Ultimately, the criterion of "tolerability" involves all aspects of the interaction of man with his environment. To cause this interaction to be "tolerable" or "compatible" is the purpose of all applied science and technology. Experience suggests that the integrated outcome of all efforts toward all goals, and the social matrix on which these outcomes impinge, need to be held within an envelope shaped by two constraints:

1. Change is inherent in the humanistic philosophy, an inescapable outcome of the application of the scientific method, and an inherent property of the natural environment irrespective of the impacts of the human culture.

2. Most if not all systematic, progressive departures in the man-environment relationship from a "steady state" have predictably catastrophic ultimate consequences.

The problem of human survival and the problem of human enjoyment may require incompatible solutions. Is it possible to achieve a steady state without foregoing the dynamism of change? At best, mankind is condemned to walk the uneasy tightrope of preserving a dynamic society in a changing environment without departing significantly from a steady state condition, without causing his environmental compatibility to become progressively degraded.

Today the steady state is threatened by countless actual and potential swings and perturbations. There is the ever-present nuclear weapon and radiation threat. There is the accumulation of non-biodegradable wastes. Electric power generation and consumption are presumed to be having an impact on the global ecology, as is the carbon dioxide from combustion of fossil fuels. Medical practice is progressively increasing the load of genetic defects that mankind transmits to following generations. Modern chemistry has multiplied to an indeterminate extent the number of mutagens, agents that like thalidomide cause birth defects, that are in widespread common use. Our preoccupation with economic modes of thought and economic criteria has permitted the wide evolution of systems of production and use of products that result in serious instabilities. The human population itself is departing alarmingly from the steady state.

In prospect are further dangers of an unsteady state: the consequences of any introduction of a pathogen into our increasingly simplified agriculture, the ecological oversimplification resulting from widespread pesticide applications on land and sea, the human oversimplification--physical, social, cultural, medical--resulting from urbanization, and innumerable possibilities of adverse impacts now unforeseen resulting from the uninhibited embracing of innovations offering desired first-order consequences.

Recognition of the disrupting effects of technology on the preservation of the uneasy balance between dynamic society and steady state of man-environment compatibility in recent years has led to proposals for a more systematic and authoritative form of technology assessment. Many students have called attention to the fact that some of the ingredients of what is now called "technology assessment" have long been operative as social functions. However, as the need has emerged for a more comprehensive, integrative function, difficulty has been encountered because of the enormous scope and uncertain priorities apparently required. Four identifiable sets of activities, more or less related, seem to be involved:

1. The forecasting of technology, necessary to warn society of future dangers and opportunities, to trigger assessments at a time when they can be most effective in terms of implementing their findings, and to enable the identification of possible trade-offs, alternatives, and solutions for assignment of priority attention.

2. The development of "social indicators," formal, quantitative, objective standards for the characterization of social status and measurement of social progress.

3. Technology assessment, itself. Here the process is recognized as an iterative one, interacting repeatedly with 1, 2, and 4. It reviews aspects of the changing conditions of man-environment compatibility, departures from the steady state, and the need for more or less dynamism at any given time.

4. The identification of national goals. This is more an educational process than a scientific achievement; it implies the establishment of an agreed-upon set of answers to the questions: "What do we need?" "What do we want?" "What do we want to become?" "What can we do?" "Should we do it?"

TECHNOLOGICAL FORECASTING

Assessment of a technology cannot be completed without making a forecast of probable trends in relevant associated technologies. Assessment of the automobile might be greatly influenced by 1. the prospect of a low-cost, safe, easily controlled helicopter, 2. the availability of a low-cost, safe, high-production, high-efficiency fuel cell, 3. a highly developed system of audio-visual and computer communications. Other possibilities of relevant future technologies might also come to mind. In making present decisions, it is often important to have an understanding of the future options. Also, what is the probable growth potential of the technology being assessed? What impacts of scale can be foreseen?

In technological forecasting, what do we want to know? The prospective producer of a technological innovation wants to know if there will be a market--if the public will be willing to pay for the innovation; he also needs to know if the innovation will work and how well. If the answers to these questions are unfavorable, he will be quick to drop the enterprise. But what if the advantage of a promising innovation is specific while the costs are generalized--for example, an electrostatic precipitator? The public can make the innovation profitable by requiring its use, but who is to decide? Who is to invest in its development? Who is to forecast such possibilities? When an existing technology is being assessed, how do these kinds of innovation become identified and considered as possible alternatives?

Still more obscure is the condition in which costs and benefits are both generalized. For example, it is easy to imagine an automatic, low-cost blower operating to disseminate air through a polluted stream so as to supplement its dissolved oxygen and improve its purity. But who should be responsible for imagining such a development? Who should develop, pay for, and assign priorities in the use of such a device?

One advantage in technological forecasting is that it enables questions to be asked early: What happens if an innovation in prospect is to be widely used? This aspect of technological forecasting appears to have been somewhat neglected. It is in the field of "normative forecasting." It asks such questions as: What is both feasible and needed? What pattern of technology should man seek? What technological developments are probable that will have high economic utility but uncertain social value?

Most published forecasts by private industry emphasize rather short-range (one-year) expectations of changes in technological practice. Or they may speculate (usually simplistically) on the economic consequences of such changes. But forecasting should go much further. If we are concerned with the trade-off between personal freedom and personal convenience, we may need to explore rather searchingly the implications of a new technology: Are we prepared to see an indefinite expansion in governmental powers--perhaps involving a great increase in teamwork of Government and industry--in order to make possible the wide and highly effective use of a significant new technology? For example, a small, weak government might have successfully blocked introduction of the automobile (as Nepal did, for years). But only a large and powerful government can successfully regulate its common use within tolerable limits. The still larger and more detailed effort needed to maximize the favorable consequences of a significant innovation, to prevent unfavorable consequences, and to integrate it effectively, painlessly, and economically, into the human environment, would call for an extension of government relationships with private industry--"interference" if you wish--far beyond anything we have yet learned to endure.

It is not really clear, however, whether or not we can actually judge in advance, in respect to some new technology, whether we want to preserve ourselves from government expansion at the cost of denying ourselves some proposed technological convenience, or to accept the government cost along with the convenience. The more complex and indirect the foreseen consequences of a prospective technology, apparently, the less likely society is to respond positively to them. Then is it acceptable to conclude that the questions are germane and important, but that nothing can be done with the answers?

In any event, one thing is clear. Forecasting is chancy at best. Although the burden of proof is on the innovator, the interdiction of a proposed new technology with clearly favorable first-order consequences is unlikely to receive social approbation, no matter how logical the interdiction may be shown to be. Only if the social costs and social gains, as well as the economic gains, can be added up and compared, and charges for these translated into economic terms will the process of filtering or screening proposed new technologies be feasible. And no way has been found to do these tasks. The nearest approach--one that is in its prescientific stage of development--is the concept of "social accounting" or "social indicators."

SOCIAL INDICATORS

The notion is that parameters of social well-being or the reverse can be identified and measured as a means of determining social progress. It is based on an analogy with economic indicators, such as those used by the Council of Economic Advisers to ascertain and forecast the economic well-being and prospects of the nation. On the basis of well-established theory, such economic indicators have been identified and measured as the gross national product, new capital formation, balance of payments, rate of economic growth, unemployment, underemployment, and the like. Unfortunately, no comparable theory has been advanced or accepted on which social indicators can be based. There are no generally accepted indicators of social progress, and every indicator thus far proposed is open to serious challenge. Some of the most devastating attacks on indicators, in fact, have been expressed by those social scientists who favor the concept in principle.

The Constitution of the United States was an early attempt at political system-building that was deliberately made flexible enough to accommodate two competing concepts of approach. Both sought the same generalized objectives of welfare, security, and justice. One stressed tranquility in a regime of law, while the other stressed liberty in a regime of "least government," with the result that these conflicting objectives are both posited in the Preamble to the Constitution. The translation of these five normative terms into operational principles has defied the best efforts of legislators, jurisprudents, and executives ever since. If they are indeed the goals of American society, how can they be measured? Is it possible to imagine some specific, concrete, objective, quantifiable factor or set of factors that can indicate the prevailing extent or change over time in the level of welfare or justice?

Perhaps more to the point, if these objectives no longer serve, if our society has outgrown them, what objectives should be placed in their stead? Have our social scientists learned, since 1787, any useful information about mankind that would enable us to chart a more positive and constructive course? For example, the emphasis in 1787 was divided between the mercantilism of Hamilton and the physiocratic leanings of Jefferson, between the centralized State of Hamilton and the decentralized democracy of Jefferson. Modern America is rich enough and developed enough to have concern for such lagging regions as the Tennessee Valley and Appalachia, and for the development of lagging regions and economies abroad. By what noneconomic parameters should progress in these efforts be measured? How do we characterize welfare or progress in a region or nation?

Or should we accept the popular verdict that judges urban crime, racial disturbances, and campus unrest as chief complaints? Should we continue to expand our efforts at education, in the light of the fact that it has traditionally absorbed the bulk of local tax dollars? (In other words, is the assumption valid that what we have done in the past should be considered to have stood the test of time and is therefore best?) What distribution of wealth and income would we regard as optimal? How is crime to be defined and measured and prevented? What are the relationships between crime, poverty, health, and education? What should be the objective of our cities--what are they for and what design would best serve this purpose? Should we establish priorities for a small number of prime social objectives or levels of attainment for a system embodying a great many objectives in accordance with some theory of balance among them?

It does not seem to me that we will be able to answer these questions until we know vastly more about mankind than we do now. Moreover, there has been a persistent but ineffectual protest for as long as I can remember--certainly since the mid-1900's--that in the social sciences our efforts have lagged behind. Today, even the practitioners in the physical sciences are willing to grant the existence of this lag, although it is not yet evident that they are willing to support a transfer of public support to correct it.

Unfortunately, the evident adverse impacts of technology on the human environment are beginning to disturb the public; they are raising questions as to the desirability of further unrestricted support of the physical sciences. If the public remains unequipped to sort out good technology from bad on the basis of sound social theory and fact, all of science may suffer. If the public, in its impatience with the failures of technology, turns against all of science, all of us will suffer. Yet I see no hope of achieving the kind of standards of social quality needed for advance testing of new technology without a great deal of effort, a much larger effort than at present being made, into society itself.

In sum, social science supports social standards. Social standards support technological forecasting. And technological forecasting is an essential component of technology assessment. For want of a nail. . . .

TECHNOLOGY ASSESSMENT

Perhaps we need a new answer to the question, what is technology? I suggest that it is every application of the cause-and-effect relationship that affects the relationship of man and his total environment. Then the assessment of technology is a determination as to man's capability of preserving a steady state under conditions resulting from the unlimited exploitation of the technology in question. It asks whether the absence of restrictions will result in an upsetting of the steady state; and if so, what restrictions are needed to restore it; and finally, by what authority and procedures the restrictions should be managed and applied.

Technology assessment is meaningless except when it is accompanied by technology forecasting and some form of social indicators. We cannot judge the merits of a technology except in the context of the alternative technological options, present and future. Moreover, technology is not assessed in terms of technological effectiveness, nor even of economic advantage, but in terms of social value. We need to apply quantitative standards that indicate objectively and comprehensively what the status of society is in relation to the results of technological changes. To paraphrase Abraham Lincoln, we cannot tell where we are going until we know where we are.

The Academy of Science's study of technology assessment performed for the Daddario Subcommittee addressed the problem of preserving technological innovation from political inhibition. I suggest it is time we recognized that science and technology are the most radical forces alive in the world today; they are imposing social changes at a rate too breakneck for society to accommodate itself to. Enthusiasm for unrestricted technological change is misplaced because it does not and never has existed. The burden of proof has always been on the innovator. Anyone who has ever spent a period of servitude in modern industry, especially in the burgeoning aerospace industry, can describe literally hundreds of technological concepts that never materialized because they were inhibited: by management decision, by the want of a customer, by policy, by inability to assemble the required talent or resources, or by diversion to other assignments. The main thrust of these inhibitions, taken together, is to encourage technologies that bring profit to the sponsoring company. It has been well-established that the profitable first-order consequences of innovation are too often accompanied by unintended, socially undesirable, second-order consequences. If technological innovation is now restricted to profitable directions and if these often are socially undesirable, the problem is one of redirecting the control of innovation by changing the inhibitions and motivations, rather than of reacting belatedly, after the fact, to those innovations that get through our present, unsatisfactory, screening process.

In short, we appear to need a better understanding of our social needs and a systematic way of determining in advance what is technologically feasible to meet these needs. Once these actions have been taken, we can then set about appraising in detail the total implications of important new technologies aimed to satisfy our public needs as these technologies emerge rather than afterwards.

It is important to stress that such a program does not imply less innovation but more. Many innovations are lost to us, or deferred, because they are not evidently profitable at the outset, or because they are not compatible with a company's product line, or because the company's ways of doing business would need to be changed to accommodate it. The conservatism of a large corporation in such matters is quite comparable with that of a Government agency committed to some particular technological course.

The point is that society is confronted with the most radical force for change in all history. For its own protection, it would seem that society should exercise some extent of purposeful control over the effects of this radical force: permitting, encouraging, even seeking out, the good, while reviewing, retarding, even inhibiting the hazardous or faulty. The assumption of cause and effect, the scientific method, has been applied to suit the interests of individual scientists and individual corporations, rather than systematically to meet the needs of society. Now that all parties agree that technology needs to be assessed in the public interest, the time is at hand for us to be scientific in our uses of the products of science. We are able to tolerate laissez faire commerce as long as it was predominantly in the public interest. When this assumption came under challenge, successful challenge, society found it necessary to impose restraints and guidelines for competition in the public interest. It seems to me that science and technology have undergone the same cycle. We should not impair the dynamic vigor and creativity of science. But we should take steps to ensure that the logic of science is fully applied. To do this requires that the cause-and-effect relationship be sought out and exploited in the determination and achievement of social goals through the systematic application of technology, in the broadest sense, beyond the inhibitions of personal interest or private profit.

The logic of applying national governmental control over economic forces was that government exists to do whatever is necessary and what the private individual is unable to do for himself. By the same token, events have abundantly shown that the individual is powerless either to control the exploitation of faulty technology or to stimulate the exploitation of technology of general benefit but no immediate profit. If these are to be done, the only recourse is to the Federal Government. In the words of Benjamin Franklin, "If we do not hang together, be assured that we will hang separately." Once, in our past history, the protesting individual could walk away from a disliked environment, to seek a new world, to carve his fortune from a wilderness. But Spaceship Earth has no more El Dorados to offer.

NATIONAL GOALS

Strictly speaking, there can be no such thing as a "scientific" or "technological" goal. The word "goal" implies that a process of evaluation, of value assignment, has been applied. To call a goal scientific or technical merely signifies that scientific or technological means are required to render feasible a politically or socially desirable outcome. To determine what a technology will do, what changes it will effect, and what consequences will flow technologically from the changes, are technical questions. To determine whether such changes are desirable is a political question. Accordingly, to seek advice from scientists or technologists as to whether goals should be pursued is as anomalous as to ask politicians to rule on the technological feasibility of achieving them.

There appear to be four broad areas in which applied research and technology can affect the man-environment relationship:

1. The exploitation of applied science and technology to resources to alter the environment in desired ways

2. The exploitation of applied science and technology to preserve and restore the physical environment of man from impairments resulting from the second-order consequences of 1

3. The application of the scientific method to the alteration of man himself, to enable him to accommodate himself more compatibly with his environment

4. The alteration of social organization to improve the compatibility of man the individual with society at large.

Decisions as to which of these areas to emphasize are a means of narrowing the choices among candidate goals. The first category was emphasized in the initial settlement of America, and has provided most of the goals sought by both private industry and government ever since.

Although the medical profession is primarily concerned with the third category, as is also the pharmacological industry, most of the gains in statistical health of mankind have been achieved in the first category. Even so, powerful Federal efforts are being supported in category three.

The application of the scientific method to the achievement of progress in the second category, especially with sponsorship by the Federal Government, is of comparatively recent origin. Pesticides came under control (in a very modest way) in 1947. The first water pollution bill was not passed until 1948. Air pollution control came under national legislation in 1967, and generalized national concern for the environment did not emerge until 1969. In the remote future is the prospect of shared international concern for the preservation and restoration of the quality of the global environment.

It is interesting that the fourth category was singled out in the Preamble of the Constitution as the principal aim of our Government. Yet, this area has proved most intractable, has seen least progress, and commands least scientific effort of all. Ways have been found to invoke economic incentives in the first three categories but rarely in the fourth. This is not to say that there has been no progress in this category. The study, Technical Information for Congress, lists an illustrative number of these.[1] But no patents, no established markets, stimulate research in this category, the results of research rarely take a tangible form that can be marketed as a product. Even in its methodology, this category presents problems: the experimental approach is largely denied to social scientists. But the human values that the applied physical sciences serve and that economic forces crudely determine are the central subject matter of the other social sciences. In the words of Edwin G. Nourse, at the time of Brookings Institution and later chairman of the President's Council of Economic Advisers:

> While scientific analysis of comparative costs and returns and investigations into the nature of the economic process occupies a pivotal place in man's effort to make a good life for himself out of the rich but reticent resources of nature, the values which come to expression in the marketplace, the preferences for certain types of goods or services, the esteem in which leisure is held, and the capacity of men to combine their productive efforts in one pattern of organization or another and in response to various kinds of incentive or motivation are matters which lie outside the field of economics as such. These contributory factors must be explored by other sciences such as

> psychology, anthropology, political science, sociology, and their handmaiden, history. We must understand the subtle complexities of human nature as well as the precise mechanistic relations of physical nature if we are to develop the national strength that grows out of productive cooperation and avoid the disruptive struggles of group, class, racial, or nationalistic warfare.

Also it was recognized many years ago that the impact of technological change generated problems that only the social sciences offered any hope of resolving. Professor William F. Ogburn of the University of Chicago declared in 1945 that "for every important mechanical invention that physical scientists make there is created a new social problem on which social scientists should work."

The "industrial revolution," caused by steam, created cities, changed agriculture from subsistence to commercial farming, built a new economic system with many new economic organizations, destroyed social classes and created new ones, redistributed wealth, revolutionized warfare, realigned the great powers, abolished the household economy, and reduced greatly the social functions of the family. The "scientific revolution" following nuclear fission of the atom may change our society and its institutions even more.

National investment in science and technology continues to be heaviest in category one, less in three, still less in two, and least of all in four. Yet, if we accept the basic assumption of the scientific method, the cause-and-effect relationship can be equally exploited in all four areas. The goals of the United States are concentrated in the first and third categories, the needs in the second and fourth. Within the past few years, the second category has received a very hopeful increase in attention and legislative attention. But the fourth category remains a lagging area. Resistance to Federal sponsorship of basic and applied research in this area continues strong. It is easy to understand the reasons for the lag and the resistance. The defects of the social science discipline are plain for all to see. Theirs is the most difficult area of all. Moreover, the very relevance of the social sciences for major social problems has intensified resistance to their development.

CONCLUDING OBSERVATIONS

In sum, national goals for science and technology do not exist. We have only social goals, just as at the time of the founding of the

Republic. Yet our social goals are normative, rather than operational. We have proposed to accomplish large technological projects, but we cannot foresee their effects on our society in advance, nor even evaluate them after the fact with any degree of objectivity. We have no standards of social progress, and see no way of developing them. It is possible for us to express technical estimates of what innovations are within early reach, but not for us to determine whether we should encourage or discourage them. There is no logical reason why we should not inhibit dangerous or injurious forms of technology; but until we develop better means of social measurement, we will continue to be incapable of assessing technology definitively in terms of what society should do about it.

So what is the answer? We can see enough dangers ahead to recognize that we must not ignore technology until we can learn to judge it properly, and to conduct our own affairs rationally on the basis of such judgments. We must do the best we can today.

But I suggest that we should make a more positive effort to explore the areas we have neglected. We should avoid being too positive and self-assured about either the opportunities or the dangers of technology. We should exploit more cautiously, and control without being extreme or arbitrary. We should recognize that direction of effort is essential, but that perfection of results is not. We should also recognize that mankind has the capability of destroying itself with its technology; we can hope that we recognize the brinks of disaster before we go over them, and learn to cooperate in backing away, step by step.

Today we do not know what sort of brave new world we want. We may never find out. But in the meantime, it is important to keep some extent of stability--the steady state--in the world we have, without losing our dynamic sense of progress and betterment.

NOTE

1. Technical Information for Congress: Report to the Subcommittee on Science Research and Astronautics, U.S. House of Representatives. 91st cong., 1st sess. (April 25, 1969).

DISCUSSION OF THE SOCIAL FUNCTION OF TECHNOLOGY ASSESSMENT

Dr. Huddle started the session with a brief summary of his paper. In his review he emphasized the problem of maintaining "a steady state without foregoing the dynamism of change." He concluded his review with this statement:

> If you are not happy with this prescription, . . . it is at least optimistic to the extent that it is the only statement for the future that I have ever encountered that did not predict catastrophe. Nor will there be catastrophe as long as we treat with equal respect those two opposing demons: the steady state and the onward and upward progress of man.

PROGRESS, GROWTH, AND THE STEADY STATE

The participants were immediately concerned with Dr. Huddle's definition of the steady state. He redefined the steady state to be "anything that is not a progressive or exponential departure from a given level or norm." Actually his primary concern was with exponential growth because "I can't imagine any significant relationship of man and his environment that can be intensified exponentially without catastrophic results." Thus Dr. Huddle's steady state did not imply an unchanging state but one which demonstrated "order-of-magnitude stability." One participant, a Government official, pointed out that fluctuations which do not involve exponential growth might still be unacceptable. For example, he noted that "if the human population of the earth were to be expected to undergo fluctuations comparable to those of lemmings or of white-tailed deer," this would indeed be considered unacceptable. "In neither case," he stated, "does the species involved become extinct; it just undergoes rather wild excursions. . . ."

Another speaker claimed that "there is an alternative to the steady state and exponential growth, and that is just gradual change." He asked whether Dr. Huddle was taking the position that "the kind of technological change that we have now, independent of whether it's rapid or slow, is undesirable." Dr. Huddle responded that our society may not be able to accept too much more technological change over the next century.

A university professor thought that the basic problem had its roots in the premium which our society places on growth:

> Our entire society and our approach to the problem of the environment and our approach to technology assessment places a premium on continuing progress, continuing growth, continuing expansion . . . . And isn't it really a problem that this kind of a goal is unattainable when we live on a planet with sharply limited resources?

He went so far as to suggest that the problem is "not merely one of reconciling the steady state with dynamism," but rather one of considering the possibility of "a little backward movement," for instance: "might it not. . .contribute tremendously to the welfare of society and to life on this planet if instead of producing more electric power we decided that maybe we ought to produce a little bit less electric power." "And do without our air conditioning?" another participant asked, clearly pointing out the discrepancy between our desire for more comfort and convenience and our desire for a clean environment. Several of those present emphasized the idea that, in the long run, continued growth is just not possible. But an economist took issue with this point and noted that there is a school of thought which holds that

> we cannot afford not to grow. There is certainly an ample amount of evidence that, when we look at social entities such as nations or firms, not growing implies death. . . . An interesting part of equilibrium analysis is in economics as well as other social sciences is not the static equilibrium that we learned years ago, but the dynamic notion of equilibrium.

What is important, he indicated, is not the fact of growth, but rather the kind of "deviations from an equilibrium path." "The growth itself is not bad," he concluded, "and I'm just sort of taken aback with the notion that it is bad."

A university professor raised the question of how one can judge the value of "the steady state or tolerable instability. . .if we really have no standards by which to measure social progress." Dr. Huddle felt that the lack of such standards should not cause us to give up our concern because "it's too easy to think of total catastrophe that renders the earth uninhabitable by systematic departures from the steady state."

Social Function of Technology Assessment

One participant was somewhat puzzled by the meaning of both the paper and its title. He asked Dr. Huddle

> Do you mean the function it [technology assessment] should serve, or the function it does serve, or how social factors affect technology assessment? I found the paper very interesting, but I don't see what it all points toward.

Dr. Huddle responded that he was concerned with the tendency to perform technology assessments as though they are engineering studies:

> what I'm suggesting is that we have to look much more to our social values in order to understand how they are put together and what the criteria of social good and social evil are. . . . We have to have some kind of methodology that society can accept and agree upon which is based upon the character of what we think man should be, should become, should move towards rather than simply saying that this is too hard to think about so let's put it aside and think about the engineering world, which we can understand.

When asked how one would go about implementing this suggestion, Dr. Huddle admitted that he had no easy answer.

GENERAL COMMENTS

A university professor wanted to know why one should place any faith in the Government to bring social concern to bear in decision-making about technological change:

> What evidence do we have that government will do such a thing? Don't we have evidence going entirely the other way--that government is contributing substantially to the very things that you're complaining about [that is, the lack of consideration of social good and evil in technology assessment]?

A Congressional Aide agreed:

> Isn't it true that the thrust of every Government program that we have now, that the whole basic concept of our law, is geared to exploitation and development? The early conservation laws, for example, really were to set the rules of the game between people who were going to exploit. . . . There was never the question of whether society has a right or obligation to say "no." If that's the fabric of our laws, how can you ask government to step in and do exactly the opposite?

A university professor felt that it was overriding importance to identify a means by which people can make decisions concerning the use of technology for the modification of the environment and of people themselves. For if we do not do this, then the use of science and technology assessment, is "part of a process which is subverting the institutions of democracy." We should really be talking about "something that is going to contribute to a perfection of the democratic process." Another participant continued in this vein:

> It seems to me that what Dr. Huddle is saying here is that we've got problems in technology and that what we've got to do is apply the technological method to ourselves and our society. That's where we've failed; we haven't been good technologists in designing this system we call society. So let's apply technology--or the cause and effect method as Dr. Huddle says--to social structure. Or let's design the technocratic method called technology assessment, which will then solve the problems of technology. Now what you're doing is using the method that has caused the problem. . .to solve it. In fact, it seems clear that what you'll do is simply create more problems. I just want to contrast that with another approach which says that what we've got to do is open up the whole decision process to public discussion, public examination, and public decision. . .so that the technology is not imposed upon the society but comes out of some kind of a deliberate process in which the public has a participation and the ability to participate in the kind of technology that is used.

Others felt that the notion of technology assessment as a "closed decision system" was misleading since even if technology assessment is "something that a small group of technically oriented people do," there is no necessity for it to be that way. Technology assessment need not be "narrowly technocratic" if social considerations are made a primary, rather than secondary, aspect of assessments. It is , it was noted, impossible to get social considerations taken into the proper account by any elite or "even by the national Government or by the Congress." But it is not incompatible with the idea of technology assessment to

> somehow strengthen or create the institutions and mechanisms in society whereby the general public and various citizens' groups of interested parties can come into the process [of assessment] and exert . . . countervailing influence against some of the other interests that are very strong at present.

The question of the openness of the technology assessment structure in this country was revived when a university professor stated that "no technology assessment is meaningful which would permit the supersonic transport plane to be built." His point was disputed by another participant who noted that

> the overwhelming assessment of the SST concluded in opposition to the building of the SST. . .the point that you have not faced up to is that what you're really suggesting is a technocratic kind of assessment. I don't like the SST. . . but I. . .sure like it a hell of a lot better than having scientists. . .making the decision as to where we will go.

A university researcher agreed though he noted that if the public as a whole becomes involved in the decision process, there will be some mistakes. "The fact that the public can decide wrongly is something that you've got to accept." He stated that a valid goal was not to have technical information control the Congress in its decision-making, but rather to have Congress make decisions with full knowledge of what the technical information is. To say that Congress should decide *only* on the basis of what the experts have told them cannot be desirable. For instance, in the case of the SST

> the information that you get from a technology assessment. . . will be only one piece of data that is before the Congress, and you can bet your last dollar that in terms of volume and money spent in lobbying it will be a trivial drop in the bucket as compared to what will be coming from the people who are building the supersonic plane.

The discussion turned to a consideration of the meaning of goals. When Dr. Huddle said that "a goal must be something recognized as desirable" the question "recognized by whom?" arose. The discussion of elites versus the general public resumed briefly and then a university researcher questioned the notion put forward in Dr. Huddle's paper that we are a "goal-oriented society." He felt that the only time we, as a society, have had goals were when we were involved in large wars. All of our goals have been individual: increasing the family's standard of living, making something of oneself, and the like. "We have never had any societal goal," he continued. Since "most of our social changes have been side effects and quite unintentional," he raised the question of "whether, in fact, a social goal is a very effective way of causing social change."

One participant noted that technology assessment is nothing new and that in fact it goes on all the time. He couldn't understand, he said, the current view that a technology assessment capability is something which must be developed. It is already developed. But a Congressional Aide responded:

> I'd be inclined to agree with you if it hadn't been that we've had catastrophes--maybe minor ones, but nonetheless they forebode other catastrophes . . . . People [in Congress] are worried about these catastrophes . . . and they don't want to allow the methods of technology assessment that are existing today to continue . . . .

Another Congressional Aide pointed out that technology assessment should be conceived of as an adversary proceeding, an argument which had been made several times in earlier seminar sessions. (See, for example, Chapter 4.) "In any society there are people who represent opposed interests" and these opposed interests must come together in an adversary process. A university researcher felt that such an adversary process would only be meaningful if we "build mechanisms. . .whereby the public, or at least citizen's groups, can get in a powerful enough position so that they can actually compete in this adversary process."

CHAPTER

8

TECHNOLOGY ASSESSMENT AND THE FOOD AND DRUG ADMINISTRATION

Alan H. Kaplan
Robert H. Becker

To those who regularly observe the activities of the United States Food and Drug Administration (FDA) it is difficult to believe that any other single agency could be charged with a more complex and frustrating role in the practical application of technology or in the assessment of that technology. The products which are subject to the agency's jurisdiction are basic and essential to the maintenance of human life--so basic that most of us must take their existence and quality for granted in the same way that we neither doubt nor question the sun's rising or the earth's ability to support our bodies.

As a consequence of a variety of dramatic and well-publicized events over the past decade particularly, we have all been made aware of the fact that such faith may not be justified. The degree to which it is justified is dependent in substantial measure upon the ability and authority of the Food and Drug Administration to operate effectively in a maze of statutory and political directives.

The scope of the FDA's problems is measured not only by calculating the vast number of products subject to its jurisdiction, but from the endless permutations and combinations of use and misuse of such products in our still relatively uncontrolled human environment.

---

*Presented March 5, 1970.

"Technological assessment," in the sense of data evaluation and decision-making based upon competing social values and evolving scientific findings is merely one way of describing the day to day activity of the Agency. From this perspective the Agency's mission of control and resolution by application of statutory criteria phrased in terms of absolutes, such as "safe" and "effective," becomes virtually unattainable. There is further indication that, from time to time, the Agency itself adds to the frustrations of the difficulties imposed upon it by taking public positions based upon absolute applications of these arbitrary statutory standards.

From a lawyer's point of view, the FDA is first and last a law enforcement agency. Its role is to assure that manufacturers and distributors of foods, drugs, medical devices, cosmetics, and hazardous substances comply with specific Federal laws and regulations relating to the quality of their products. Where this is accomplished through voluntary compliance programs rather than criminal prosecution it is simply a use of the carrot in lieu of the stick. Where the agency issues regulations prescribing or proscribing particular forms of foods or drugs, it must rely upon the courts to require adherence to such regulations, and the courts must find the regulations authorized and justified by the statute. Where the FDA undertakes independent research or gathers information about a product or its effects, it is, ultimately, investigating to determine whether a case should and can be made against someone. The evaluation of existing foods, drugs, or cosmetics, and the establishment of requirements for the premarketing clearance of new products may well be a kind of classic situation for examining the general concepts relevant to technology assessment in a regulatory setup under the American legal system.

To fulfill its responsibility the agency should do the following:

1. Know and apply the latest and best methods of the technology available to evaluate the products under its jurisdiction
2. Require the use of such methods by applicants seeking premarket clearance
3. Evaluate the data pertaining to those products through the application of the public policy presumably established by the relevant statutes.

Obviously, the agency not only must be aware of developments in the technology which can be utilized in evaluating the products subject to its jurisdiction but also must be in the forefront in applying and requiring the application of such methods. Unfortunately, one of the facts of human nature seems to be that when there is a "big brother" who tells us what we must do to satisfy him, that level of performance often becomes the maximum as well as the minimum which will be done by some. This is not to say that industry scientists are not constantly attempting to develop and apply new methodology, but simply that it would be foolish to assume that many persons in any industry would hold up applying for clearance to market a new product after meeting established Government requirements because of an awareness of yet another test that could be applied to the product.

But there is another side to this also. Because of the political exposure which, in the past decade, has become an important fact in the life of the FDA, there is always a danger that the Agency, through its in-residence reviewers, will find it safer to require additional testing, not because its results are reasonably expected to supply scientifically meaningful or useful data, but simply because the test can be done, and, some day, someone may ask why it wasn't done. After the thalidomide tragedy, for instance, tetragenicity studies in laboratory animals became a political necessity although the actual meaning of either positive or negative results was and still is open to considerable scientific question.

There is nothing novel in the observation that decisions involving the assessment of technology necessarily involve the application of social values to scientific data. As applied by a regulatory agency it would seem obvious that this includes two limitations on the scope of action to be taken: the nature of the data available and the social values which are allowed, by statute, to be considered. For good or for bad, the Agency is not legally justified in applying factors or considerations outside the scope of its statutory charges.

There are, of course, many cases in which the decisions are relatively easy. However, when the available technology cannot, with reasonable effort, supply data necessary for a particular decision, the decision cannot be made on a scientific basis. This is akin to the mathematical principle, that conclusions cannot be stated in terms any finer than the system of measurement available for supplying data for those decisions. Yet there are times when decisions cannot be deferred or when deferment is a decision. When data are inadequate and the technology cannot reasonably supply more adequate data, the decision must still be made and, whether the product is allowed or not, the risks are often substantial. Certainly, the withholding of an effective drug because of a demand for additional evidence can be as costly in human suffering as permitting marketing of an unsafe or ineffective one. The bases for such decisions are quite obviously political in the finest sense of the word and have become political in its most common sense as well.

What policies and objectives is the Food and Drug Administration to pursue and apply in making these decisions? Here, to lawyers, is the very heart of the question. Our legal system is founded upon the maxim that we are a government of laws, not men. The imposition of controls is to be kept as free as possible from the whims and caprice of individuals. It is now many years since serious students have believed that such freedom is possible, but administrative law is a constant struggle to avoid arbitrary rule by requiring adherence to some standard, however vague. This is no less true in the food and drug field than any other.

It is noteworthy that for the most part the criteria provided by the statutes under which the Food and Drug Administration operates are narrow and negative: New drugs may not be marketed unless they have been shown to be safe and effective. Food additives may not be used until proven safe. Labeling may not be false or misleading. The Agency's mission is to protect us from. . . . It is by these standards that the FDA must judge, and by which its decisions are publicly judged. If a food additive is alleged to be "unsafe," the statute does not allow distinctions between a product which could double the supply of an important food, and one which would only slightly reduce a manufacturer's costs of production. This might be reasonable if "safe" were not so abstract and illusory a concept. But when we arrive at a situation in which cyclamates are "unsafe" in foods because they caused tumors in rats in one study--with the result that many foods containing cyclamates are now illegal regardless of how they may be labeled, but cyclamate-containing table sweeteners, labeled for use by diabetics and obese persons under physician's care, are being approved as safe and effective for use as drugs--it is fairly obvious that the law has become a fudging ass.

We do not mean to criticize the Secretary's attempt to make some sense out of a difficult regulatory situation, but we are only commenting upon the law, which too often requires absurd gyrations to achieve even a modicum of reason in its results, and asking why.

The two most publicized and criticized areas of FDA activity are the premarketing clearance of new drugs, and the premarketing clearance of food additives. These also provide informative illustrations for the applicability of statutory standards to technology assessment.

NEW DRUGS

The heart of the Food and Drug Administration's regulatory authority over drugs, and one which is assuming greater significance from day to day, is the statutory prohibition against the interstate shipment of a new drug unless there has first been an affirmative approval for the marketing of that specific drug. The process for obtaining such approval involves first, the collection of sufficient animal data relating to the drug's safety and activity to justify the commencement of human trials; second, a series of human clinical tests progressing from trial in a limited number of volunteers for the purpose of determining as much as possible the drug's pharmacological activity in humans, through large-scale therapeutic trials in patients suffering from specific conditions against which the drug is thought to be of potential utility. All human trials are required to be done under protocols submitted to the Food and Drug Administration at least concurrent with the commencement of such trials. Finally, when the sponsor of the drug believes that sufficient data have been developed, a "New Drug Application" is submitted requesting approval to market the drug commercially.

It may be instructive to analyze in more detail this brief summary of our drug clearance system for the many pertinent problems of technology assessment inherent in each step and to examine their resolution under present procedures.

Initially, the whole process is dependent upon the adequacy of the animal model to determine the effects of drugs in humans. Most pharmaceutical research screening is now done in lower animals. This substantially reduces the prospects of discovering a compound which would have a desired quality in man, but not in the test species. Drugs which may be effective in man as well as other species of animals may and probably are being bypassed because they are toxic at the initial screening level. The time and expense involved in determining whether that toxicity is specific only to the screening species are simply too great to devote to any compound. The potential loss to society incident to that inability is illustrated by realization of the fact that insulin is probably one compound which would probably not have passed animal toxicity tests in the dogs now commonly used for that purpose. Similarly, drugs which are apparently safe in the species tested may prove to have other dangerous effects in humans--the classic illustration being thalidomide. But the technology, however imperfect, must be recognized as having brought about a therapeutic revolution with remarkably minimal, and up to now, certainly acceptable, fallout in terms of harm to the human population.

There is now much talk about the slow-down in the development of new drugs during the past decade. Undoubtedly, the incidence of significant new advances has slowed. But the "blame," if any there be, may be attributable largely to the fact that we have just about used up the comparatively easy finds discoverable through use of the old screening testing technology, and that for the future we must develop and utilize new techniques to discover and develop new compounds suitable for pharmaceutical testing as well as to improve testing techniques to satisfy the scrutiny being given by the FDA to the more extensive and intensive evaluations now being required. Obviously the more intensive and extensive the testing required and evaluations applied before a drug is marketed, the fewer are the prospective drugs that will be able to be investigated. Drug investigational resources in both personnel and funds are limited.

This relationship of "tighter testing--fewer drugs" then results in a phenomenon which may or may not be consciously considered by those establishing the test standards. In the interest of safety, there is no doubt that extensive animal and human testing should be done on any compound before it is considered for human use. But each time we increase the margin of safety, the number of compounds which can be investigated by the available resources decreases. How does a regulatory agency measure imponderables such as these? Is the increased safety provided by extending animal toxicity testing from three or six to twelve or eighteen months more valuable than would be the application of the resources used for such purposes to screen another compound? A small question in assessing the FDA's assessment of technology, but one on which responsible toxicologists strongly differ with FDA. Yet, to our knowledge, no one has attempted to determine how often an additional year's testing has provided significant new information, nor can we predict what percent of compounds left untested because of these requirements would have become new miracle drugs.

In our experience these are not the kinds of questions which have been considered. Instead, as suggested earlier, there is a tendency to fix upon the unprovable absolute: a product must be shown to be "safe" no matter how impossible, impracticable, or illusory that concept is in absolute terms. The other considerations, e.g., the allocation of resources or the delays in the availability of useful new products, appear to be sometimes ignored in the attempts to reach politically acceptable assurances of safety.

The final step in clearance, the approval of the New Drug Application, is itself a case study in technology assessment. The "sponsor" or manufacturer's medical and scientific staffs have determined that the data are sufficient to justify marketing. The FDA's staff is now faced with the decision whether it should or should not allow this new, probably highly potent, compound to be administered to tens or hundreds or thousands of humans. Even assuming relatively unambiguous data as to the effect of the product in the pre-marketing trials, how is a product judged safe? If a product cures or causes remission in Parkinsonism, but has the potentiality to cause gastro-intestinal bleeding, adverse personality changes, or other ill effects, in a finite percentage of patients, at what point has it been tried on a sufficient number of patients in controlled studies so that there is sufficient assurance that its actions are well enough known to release for use by all persons suffering from this disease. When is it to be allowed on the market so that the physician and patient may be permitted to decide if it should be used, or must that decision still be made for all by the FDA?

These are obviously political questions of the highest order. Since they cannot be publicly debated for each daily decision, the FDA has to look to its statutes for guidance as to the policies to be applied. And, as is so often the case, the law's answer is inadequate, if not ludicrous.

The New Drug Application is to be approved if the drug is shown to be "safe" and if "adequate and well-controlled investigations" show that the drug has the effect that it purports to have.

What is safe? By their nature drugs are designed to actively interfere with ongoing biological processes. Because of the technology employed, the compounds tested as drugs are chosen because they have general rather than specific activity. Both by biologic logic and experience we know that the harmless drug is an impotent drug. Very early in the administration of this law the Government recognized that safety could not be applied as an absolute term. For instance, in testimony before the Senate Antitrust & Monopoly Subcommittee George Larrick, then Commissioner of Food and Drugs, stated:

> The safety of a drug cannot be considered as an absolute value to be determined with a computing machine or measuring stick. All potent drugs, old and new, have a potentiality for harm.
>
> If we were restricted in our decision on whether to clear a new drug for distribution solely to those that are harmless, then we would have none of the life-saving drugs like the sulfa drugs, corticosteroids, tranquilizers, etc. So the medical officer and the other scientists must evaluate safety in the light of the good inherent in the drug and balance it against the hazards. If the good in saving lives or alleviating suffering clearly outweighs

> the hazards, he will permit the NDA [New Drug Application] to become effective under labeling which clearly sets forth the hazards of the drug, the contra-indicated conditions, the side effects, cautions to be observed, warnings, and directions for use which reduce to a minimum the danger inherent in the use of the drug.[1]

So the expected benefits from the use of the drug have to be weighed against its known and expected adverse effects. The law has delegated this responsibility to the FDA, which evaluates the data and allows or disallows marketing on the basis of finding the drug safe or unsafe, effective or ineffective.

In making this decision for all of us, the FDA doctor, interestingly, is not allowed to consider "comparative effectiveness," that is, the existence of another drug, either more or less effective than the new drug, in accepting or rejecting the new drug. The Congress, quite rightly, we believe, decided that today's technology is not adequate to require or allow any bureaucracy to limit the available treatment for a disease to one or another drug which it believes best. The variation among patients, the differences in the natural history of the same disease under different circumstances, and the impossibility of developing absolutely accurate data on which to make such judgments, clearly require that all "safe and effective" drugs be allowed to be marketed. Comparative data, if pertinent, can be left to labeling.

What constitutes a "safe and effective" drug in a legal sense has until fairly recently been a moot question. There is, of course, a broad area of agreement as to what constitutes an acceptable risk, and in those cases, where there might have been a difference of opinion, sponsors of new drugs have generally been able to read the handwriting on the FDA medical officer's wall fairly early in the investigational process and, rather than complete the lengthy and expensive process only to face an almost certainly losing fight, have gone on to the next-promising compound.

Now, however, pending legal cases involving widely prescribed combination drugs are likely to raise some of the more basic unresolved questions as to the meaning of "safe" and "effective." How much more effective does a combination drug have to be than either of its components in order to justify its marketing? in how many cases? and how many adverse reactions can be added by the second component and still have the drug be called safe? How far these answers will carry over into the meaning of these terms as applied to single-entity drugs is an open question.

All of this poses for us one of the ultimate questions of technology assessment in this area: has the technology for evaluation of drug effectiveness and safety or the criteria for the assessment of that data reached a point where they are adequate to ask anyone to make yes-no judgments in all cases? Should decisions concerning the use of drugs with what some might consider undesirable toxicity-effectiveness balances be left to the physician or even his patients so long as accurate and truthful labeling requirements are met? Or should the fact that FDA doctors or the FDA Commissioner would not use a drug or practice medicine in a particular way constitute adequate grounds for withholding or suspending approval of a new drug?

In some cases the Commissioner has shown advisable restraint. At times, where the balance was unproven and, for the moment unprovable, physician and patient were allowed to assess and assume the known risks. An example of this is found in drugs used to lower cholesterol levels. Although there is not yet a proven cause and effect relationship between high cholesterol levels and heart disease, many physicians and informed lay persons are of the view that the importance of lower serum cholesterol levels has been sufficiently indicated for them to act upon it. This, in some cases, includes assuming certain risks associated with cholesterol-lowering drugs. It would have been quite possible for the FDA to have required conclusive studies that cholesterol level-lowering drugs actually reduce the incidence of heart disease before allowing them to be marketed. This was not done. We think this a wise decision. When the relationship of cholesterol levels to heart disease is finally established appropriate changes can if necessary be made. In the meantime the physician and his patient are at least in a position where they are permitted to make their decision on the basis of all of the available information.

Perhaps "the Pill" constitutes the most notable illustration of the difficulties confronting the FDA in its application of technology assessment from both the scientific and social aspects. The Pill is the only "drug" taken on a continuing and sustained regimen by normal healthy individuals. What benefits must be available under such circumstances to offset the risks that are known to exist? Utilizing the traditional "benefit to risk" scale of evaluation, it might appear to be a relatively easy decision (to some at least) to restrict marketing of the drug under labeling which recommended limiting its prescription to women whose life would be placed in substantial physical jeopardy through pregnancy. The reciprocal of that decision would, of course, involve recommending the withholding of the drug from women physically capable of normal childbirth. But the decision involves much more than the weighing of risks to the physical safety of the individual woman. Psychological, economic, and indeed social and moral considerations are all involved. Certainly, because of the existence of known hazards with the Pill and the still unresolved uncertainty of its safety in long-term use, a decision could have been at least arguably legally justified to deny approval of new drug applications. In fact, it has even been argued that the Pill has been made available in contravention of the statutory standards. But the decision that has been reached, that of "freedom of choice" on the part of the physician and layman, subject to the requirement of "full disclosure" concerning the known propensities of the drug appears to us to have been fully justified.

What we suggest ultimately is a policy under which the FDA should refrain from decisions which can be avoided by allowing the decision to be made by other competent professionals closer to individual situations. Requiring truthful disclosure of the known effects and possible damage caused by a product should in many borderline cases obviate the need for decisions which, if wrong, would deprive rather than protect. It would be unfortunate if the statutes either prohibited, or were interpreted to prohibit, consideration of those factors other than proven safety and effectiveness, which would allow this result.

FOODS

Similar problems concerning the safety of food and food additives as well as pesticide residues are now also beginning to surface in the public political arena.

Cyclamates, monosodium glutamate, and DDT are, fortunately or unfortunately, familiar to all of us both intellectually and gastronomically.

A beautiful illustration of the difficulties inherent in decision-making where the technology has not provided adequate data and where the law provides insufficient guidance for technological assessment is illustrated by the recent controversies over the use of chickens with avian leukosis. The scientific situation was aptly stated by Ralph Nader as follows:

> We have heard the statement by some in the Department of Agriculture that there is no evidence that avian leukosis contributes to or has an effect on human diseases. . . . There is every justification for saying the opposite. . . . There is no solid evidence to prove the matter either way.

If our policy is "proven safety," with the burden assumed by the persons marketing the product, then the decision should be, as it was, that these chickens should not be marketed. According to Representative Sullivan, only a little over 1 per cent of inspected poultry is rejected for leukosis. We would have serious doubt as to the reasonableness of that decision if, however, as Mr. Nader suggests, avian leukosis "afflicts from 40 per cent to perhaps all the chickens grown in this country." Would the elimination of an unknown but possible potential health hazard justify the elimination of this food and the industry producing it?

HEW'S answer to such a situation was suggested by Secretary Finch's statement relating to the continued, although restricted, use of DDT:

> If the Delaney Amendment, as it is now written, were to be enforced for pesticide residues, it would convert us to a nation of vegetarians. Much of our red meat, many dairy products, some eggs, fowl and fish . . . all parts of basic food groups deemed necessary to a balanced diet . . . would be outlawed because of very small pesticide residues from the ecological chain.

Cyclamates, albeit in substantially higher quantities, were, as previously mentioned, ordered removed from general purpose foods and limited to use in drugs to be used by diabetics or weight reducers under a physician's care.

If a thread can be found through these regulatory assessments, it is that more is needed than the application of proven "safety" as the ultimate standard by which to assess developments. Yet, here again our statutes leave much to be desired.

The Federal Food, Drug, and Cosmetic Act provides that a proposal for a food additive regulation shall be denied if it "fails to establish that the proposed use of the food additive, under the conditions of use to be specified in the regulation, will be safe." In one of the most absolute and specific cases of Congressional technological assessment, the act further provides that no additive shall be deemed safe if it induces cancer, when ingested, in any animal. The Congress thereby legislatively enacted the nonexistence of a threshold level for carcinogenisis and the applicability of animal carcinogenicity testing to humans.

Thus, the statute requires an unbalanced, undefined, and often unattainable "safety" for marketing a food additive. Given a toxicity profile which, in the FDA view requires or justifies a finding that safety has not been shown for a substance that reduces baking time for bread, the same toxicity profile should, legally, prohibit the use of a residue-leaving growth promotant which doubles the acreage yield of wheat. Still, despite the apparent rigidity of the law, we do not for one moment believe that such a result would actually follow. These other factors, the effect on the food supply, the effect on nutrition, and even, at least under some administrations, costs, will affect decisions as to "safety."

As lawyers, however, the probability of acceptable results through the misapplication of laws offends us. It would offend us even more if a client felt that the overall balance of relevant factors which should be considered fell on the side of allowing the use of the product, and that could be proven if such relevant factors could legally be considered, but the FDA decided otherwise, and was then required to defend its decision only on the grounds of "safety."

While dealing with technology assessment and foods, we should at least mention the pending proceedings relating to the establishment of standards and labeling requirements for all foods to which vitamins and minerals have been added. At this time, and for the past twenty months, there has been a continuing administrative proceeding convened for the ostensible purpose of testing the reasonableness of FDA regulations dealing with foods "for special dietary uses." The regulations underlying the proceeding were published by the Agency in 1966, four years after the issuance of proposed regulations which themselves generated substantial comment and concern. The 1966 regulations revised substantially the 1962 proposal and additionally introduced many novel and controversial concepts imposing limitations upon food formulations and labeling concerning their vitamin and mineral properties. The legality as well as the scientific and the social rationale of these concepts are presently, and for years to come may remain, the subject of substantial and costly dispute. The assessment of the state of scientific knowledge covering vitamin and mineral supplementation for our food supply is proceeding by means of testimony by and cross examination subject to legal rules of evidence. Yet, it is likely that much, if not all, of this dispute could have been avoided had the FDA consulted with and sought better technical advice from recognized nutritional authorities concerning appropriate standards to be embraced in the regulations. It is more than likely that a consensus would have been achieved acceptable to the affected segments of industry and of meaningful utility and benefit to the interested consumer. Additionally, in promulgating its regulations which, in significant aspects, were predicated upon a variety of administratively made assumptions concerning consumer attitudes and understanding, the FDA failed to utilize the more reliable and valid techniques of consumer surveys which are available for assessing such attitudes and understanding. It was largely because of the failure of the Agency to apply appropriate assessment techniques that this unfortunate hearing came into being. But at least in one sense the hearing is fortunate--it has constituted an open, if inadequate forum at which the alleged arbitrariness of Governmental action can be aired.

CONCLUSION

We have tried to develop the importance of statutory standards in technology assessment in a regulatory setting. They are the only defense to arbitrary actions as well as the guide for administrative action.

We recognize difficulties inherent in the formulation of meaningful standards for the kinds of decision-making involved. For foods, the words of the statute are in many respects inadequate. For drugs, while there is more flexibility in language, a statutory recognition of the various other objectives involved would be useful to both the regulator and the regulated.

NOTE

1. U.S. Senate, Committee on the Judiciary, Subcommittee on Antitrust and Monopoly, Hearings pursuant to Resolution 238, 86th Cong., 2nd sess. (June 3, 1969).

DISCUSSION OF TECHNOLOGY ASSESSMENT AND THE FOOD AND DRUG ADMINISTRATION

Mr. Kaplan opened the evening's discussion with a brief summary of what he felt were the important issues raised in the paper. He emphasized that he and Mr. Becker should not be considered as objective observers; as food and drug lawyers "our professional life has been built around the everyday activities of the Food and Drug Administration."

The FDA is a regulatory agency with a negative mission. That mission is to "protect the public health of the nation as it may be impaired by foods, drugs," and certain other products. The agency's mission is not to develop new and useful products or to encourage the discovery of new drugs. It is to protect the public against the marketing of potentially harmful drugs. Mr. Kaplan felt "that this is an unfortunate mission" and that perhaps a different focus, a more positive one, might make the agency more effective.

He noted that the FDA itself has sought to change its mission. One example of this is the warning which the FDA requires to be inserted in all packages of birth control pills. For many years the FDA opposed the notion that drug manufacturers should supply information directly to a consumer of a prescription drug. This was felt to be the function of the physician rather than the manufacturer. But now, apparently because of an FDA view that information being supplied to the patient in the case of oral contraceptives is insufficient, the FDA has "embarked on a new program to provide direct information to the lay consumer." Mr. Kaplan claimed that there is considerable question as to whether the statute under which the FDA functions gives the FDA the authority to carry out this program.

The FDA, Mr. Kaplan pointed out, operates under a statue which is its charter and which describes the objectives and obligations of the agency. In recent years the agency, "because of the strict limitations imposed upon it by the statute," has been forced to go beyond the scope of its charter and to apply policies broader than those which Congress has set forth in the statute. The FDA is charged with assuring the "safety and efficacy of drugs"; it has gone beyond this to consider other social costs and benefits of new drugs. Mr. Kaplan noted that

> . . . we don't criticize the application of these policies . . . as a question of social necessity. We do criticize the application of these policies as a matter of legal propriety. The agency is, first and foremost, a law enforcement agency. It takes its charter from the statute. It is an agency of supposedly limited power and yet it arrogates unto itself a power of regulation beyond the scope of its legislative charter. Again, we do not question the necessity of the agency coming forth with social policies and using means beyond those provided by statute. But we do question whether any agency can and should properly undertake to take into consideration in its regulatory policies concepts that are beyond the statute.

Mr. Becker emphasized that in any technology assessment social standards must be applied in the evaluation of the consequences of the use of a technology. In the case of the FDA, the "standards are very often amorphous or not formulated" at all. He repeated that it was of vital importance that FDA, as a law enforcement agency, have a clear mandate to consider a broader range of social factors in its assessment. It is important to "find some way of. . . giving the administrator [of FDA] more guidance as to what he should be doing." He noted that the "only priority that we have set is a kind of emotional one, that we want safety." Although there is nothing wrong with wanting safety there are clearly other factors which must be considered.

Mr. Kaplan noted that when FDA goes beyond its actual authority, as it often does, the courts generally uphold its action on the basis of broad public policy. Thus, although a lawyer might claim that the FDA cannot consider, under the law, factors other than safety and efficacy: "there is nothing that can be used to show a lawyer wrong so much as the fact that the courts say that what he says can't be done." Actually, Mr. Becker pointed out, the law is what the court says it is regardless of the views of food and drug lawyers. But since the FDA does, in fact, apply broader social standards than a literal reading of the statute would seem to indicate, it would be desirable to have embodied in the statute "<u>all</u> of the kinds of considerations" that are necessary to arrive at decisions on the use of foods and drugs.

A BROADER MANDATE FOR FDA

A university professor suggested that such an improvement in the statute might result if "we eliminated the standards of safe and effective with respect to drugs and substituted. . . , as we have in other administrative agencies, the general standard of. . . the public interest." Mr. Kaplan noted that the agency does apply this more general standard; "they use the public interest as their guiding light." Mr. Becker felt that the suggestion of using "the public interest" explicitly in the statute was a good one because it would enable FDA to apply broad considerations without depending upon the courts to allow an extension of the provisions of the statute. In fact the use of the term "public interest" would have the benefit of focusing the thoughts of the regulators on the broad aspects of drug use rather than only on safety and efficacy.

But a Congressional Aide was quite concerned with the potential implications of such a change in the statute:

> It seems to me that is a fine proposal in a Utopian society. But in fact we are so ambivalent about what our social purposes are that you may be removing a very valuable safeguard. What shines forth. . . is that we

> don't have any consensus as to national priority or policy and that we are extremely ambivalent as to what it is we want the FDA to do.

When laws are to be administered in the public interest or for the public benefit, someone must ask, "who is the public?" He continued: "The minute you tell an agency to do something in the public benefit, it is immediately besieged, I think, with people who have a clear idea of what the public benefit is." A university professor agreed. Public interest "is a vague phrase and everyone associates himself with it." Another professor noted that "safe" is just as vague and that "the courts will, in fact, interpret safe as being. . . in the public interest."

Mr. Kaplan agreed that "public interest" is a vague concept but said that "general words would satisfy us because they give a greater latitude to the agency." To this, another participant responded that:

> . . . you are really bumping for "public interest" rather than a simple matter of safety and efficacy because you say it is so difficult to describe safety and efficacy. But when you try to define "public interest" as against defining safety and efficacy, it [the difficulty] is compounded. . . . And it's clear that as a member of a law firm you're delighted to have as fluid and flexible a matter as the public interest to worry about.

But Mr. Becker held that the real concern was to open the path to the discussion, in court, of all the factors other than safety and efficacy which make up the public interest. To his statement that he wanted the statute to direct the FDA to use, and the courts to decide upon, broader issues than merely safety, a participant observed "To paraphrase Pogo, I have seen the enemy and they are you." This participant felt that the use of a public interest standard was merely a ploy to allow drug manufacturers to circumvent the safety standards now in effect.

At least one of those present felt that the application of a public interest standard would, in reality, make it more difficult to put new drugs on the market. He felt that it was important that new drugs be used and that the drug industry not be stifled. The public interest standard would be an unnecessary constraint on the market.

But others made clear that they felt that a broadening of the statute would make it easier to put new drugs on the market. One participant said: "I'm guessing that allowing the Administration to overtly as well as covertly consider factors other than the negative factors of safety . . . would make it, if anything, freer to consider the pluses and would allow them [new drugs] to get through more quickly and more easily."

Another noted:

> . . . there is a great difference, I think, between enabling an agency to arrive at a conclusion that a product is safe and . . . [having it] arrive at a conclusion that the marketing of a product would be in the public interest. I think that there is a greater psychological latitude given to the agency there so the obstacles that might be raised to the marketing of a drug might be much less under the change in vocabulary.

One problem which arose several times during the course of the discussion was the lack of explicit priorities in FDA policy. Thus, sometimes FDA seems to put most of its emphasis on regulation of food items; at other times its primary interest is in drugs. But a Government official was surprised that this should be a problem. "Any Federal agency has a statement of priorities called its budget" and the budget must be defended before the Congress each year. Another participant, however, pointed out that the system of Congressional review is currently ineffective. Of the FDA appropriations hearings he stated that "it's like driving through a small town: if you don't slow down, you don't see it." In the view of a Government official this was a fault of the Congress; the Congress is "where the public interest is supposed to get cranked in."

There was brief consideration of a point raised in the opening remarks which indicated that one of the FDA's stumbling blocks was its negative outlook and its lack of authority to promote as well as regulate the introduction of new drugs. A university researcher put this question to the speakers:

> [Your statements contain] the notion that the negative direction of the FDA somehow hampers its effective performance of the overall protection of the public and that somehow a statute that allows the FDA to perform a more positive role with respect to drugs in addition might be advantageous. I would just like to ask whether you really feel that a single agency can both promote and regulate a technology?

Mr. Becker responded that he thought it was absolutely necessary for the agency to do both since, presently, in making regulatory decisions, the agency is, in effect, making decisions concerning the direction of drug development and use without taking into account the question of which direction is most desirable. This can lead to unfortunate results, such as blocking potentially promising lines of development. He felt that "promoting the yes's seems to be at least as important as giving the no's."

DRUG INFORMATION AND THE CONSUMER

There was considerable interest in the question of whether or not drug manufacturers should provide information about the effects of drugs directly to the consumer. A university professor wanted to know the basis of the FDA policy against such provision of information. Mr. Kaplan responded that it has been the FDA's position that "the information is properly directed at the physician. The patient should get his information with respect to the drug from the physician, not from the manufacturer." One participant noted:

> . . . many physicians prefer to keep the identity of the product that they are giving to the patient unknown. Now I think that modern medicine is moving in the direction of identifying prescriptions . . . but there is certainly a backlog of older physicians who are not accustomed to doing this. . . . The idea that a patient will receive, directly through the leaflet inserted in the package, not only the identity of the drug but the identity of all the side reactions seems to me to be opening up a Pandora's box of problems for the practice of medicine.

In the case of the leaflet inserted in packages of the birth control pill, Mr. Kaplan noted, "the leaflet was drafted to work on the premise that it is merely a reminder of what your doctor has told you." In the past "it has been the obligation of the medical profession to advise the patient of the effects of the particular drug prescribed," and the FDA has sought to keep the drug companies out of the business of providing this kind of information. The policy of including a leaflet in packages of drugs is, in effect, a criticism of the way that the medical profession has been meeting its obligation.

A university professor raised the question: "Is the patient competent to be the judge" which medicine is best for him? A Congressional Aide responded by stating that the decision to provide some information directly to the consumer

> . . . reflects a growing sophistication and lack of trust on the part of the individual consumer. The average person is becoming vastly more sophisticated about what is being done to him and is raising vastly sharper questions about it. And the medical profession _is_ being criticized and rightly so in the eyes of many people, myself included. You can't just look at it just for its implications and discomfort to the medical profession. This . . . makes me a little bit more comfortable about what people are going to be doing to me as doctors.

When it was pointed out that consumers often ignore information about potential detrimental side effects, as in the case of cigarettes, another participant noted that while improved labeling and information dissemination will not help all consumers, they will certainly work

> . . . to the benefit of the interested consumer, the one who is willing to take heed of the knowledge that is made public. And I think that there is a distinction to be drawn between the consuming public at large and the interested consuming public.

A university professor, however, noted the FDA has the responsibility for safeguarding the public at large and not just interested consumers. This speaker did not suggest an alternative to providing information to the consumer, but did indicate that if only a part of the public benefits from the information, then some other mechanism might be necessary.

NEW DRUGS AND TECHNOLOGY ASSESSMENT

A Government official asked whether this country has the capacity to perform effective assessment of new drugs. One participant felt that there are just not enough people who are "really good" to keep up with the rate of introduction of new drugs. This leads to a question of priorities: if everything cannot be assessed completely then which areas should be ignored and which should be stressed. At present, he felt, no one is setting these priorities, so the FDA is left to make priority decisions without any policy guidelines.

One Government official was concerned that "assessment was limiting the technology." With the huge number of drugs being introduced, it becomes impossible to perform assessments quickly enough to keep up with the rate of development. Thus manufacturers are forced to slow down the development of new drugs. But another participant took issue with this line of thinking. He felt that

> . . . the assumption of this discussion all along has been that the technology proceeds apace and that the FDA has got to do the best it can to keep up with it. . . . As a completely heretical and obviously politically impossible suggestion, I wonder if it wouldn't be possible to put a stopper to technology a little bit and allow the process of assessment . . . to catch up . . . with technology.

When asked by a university researcher, "Would you give up the search for a cure for cancer because some of these drugs might be dangerous," this speaker replied that this was a matter which should be thought about and discussed. He wondered whether, on balance, we might not be better off without some effective drugs if that would also mean that we would not be exposed to dangerous ones.

Some of those present felt that the FDA was not capable of performing assessments on any but the most narrow basis. A university professor felt that:

> . . . FDA without question has the ability to determine the character of a particular drug or food additive. . . . But can they . . . or should they make the ultimate decision on the social consequences or trade-offs connected with a drug?

A Government official noted another problem which arises when the FDA seeks to assess new drugs: "The FDA doesn't do all of the clinical tests: it expects the drug industry to do it themselves, and every now and then the drug industry completely fakes out the FDA and runs in phony evidence. . . ." But the FDA has little recourse but to count on others to perform some of the testing: "The agencies--as government agencies now seem to operate--are very unlikely to have the in-house competence to really carry out . . . long-term studies." A Congressional Aide was concerned about the way that the adversary process was being applied to the assessment of drugs:

> You have the drug industry doing a technological assessment designed to con the technological assessors of the FDA. And the technological assessors of the FDA are carefully watching the carrying out of the task to be sure that the drug industry is honest or complete.

He doubted that this was the most reliable, effective, or efficient way to perform a technology assessment.

CHAPTER

9

TECHNOLOGY ASSESSMENT OR TECHNOLOGY HARASSMENT: THE ATTACKS ON SCIENCE AND TECHNOLOGY

Leon Green, Jr.

INTRODUCTION

Although the subject of "technology assessment" has only recently come into high vogue, the process has been going on a long time, possibly since the evolution of Man the Toolmaker from his simian forebears. According to Judeo-Christian mythology, the earliest recorded example involved divine assessment of mundane technology used to produce urban sprawl and high-rise monstrosity:

> And it came to pass, as they journeyed from the East, that they found a plain in the land of Shinar; and they dwelt there.
>
> And they said one to another, Go to, let us make brick, and burn them thoroughly. And they had brick for stone, and slime had they for morter. And they said, Go to, let us build us a city and a tower, whose top may reach unto heaven. . . .
>
> And the Lord came down to see the city and the tower, which the children of men builded. And the Lord said, Behold . . . this they begin to do: and now nothing will be restrained from them, which they have imagined to do.
>
> . . . .

*Presented March 26, 1970.

> So the Lord scattered them abroad from thence upon the face of all the earth: and they left off to build the city.[1]

In Greek mythology, mankind's exploitation of Prometheus' gift of fire was viewed with similar disfavor by Zeus. The Prophet Mohammed warned: "Every innovation is an error; every error leads to Hell Fire." Subsequently the assessment process devolved to mortal, if less than friendly, authority vested in a series of pharaohs, kings, and other elite groups. Fearing "technological obsolescence" of the metals industry, Emperor Tiberius suppressed an advanced technique of glassmaking. The sixteenth century popes looked askance at the printing machine of Gutenberg, which disseminated heresy at an unprecedented rate. Even Erasmus, scholar, writer, and editor, was nervous about the new device which permitted radical ideas to reach unscholarly eyes. Since those times the bases of assessment authority have been increased, and indeed now the process shows hopeful signs of becoming democratic.

How best to achieve this goal has been belabored of late by extensive governmental and public discussion in the form of seminars,[2,3] reports of the National Academies of Sciences and of Engineering,[4,5] recent Congressional hearings,[6] research conferences,[7] scientific symposia and forums,[8,9] and other learned discourse. The subject of all this exposition has been the rational management of technology assessment in the various branches of the Government (even including a proposed "evaluative" branch) and its regulatory agencies, and the information systems required for such management. The present paper will address not how the assessment process ideally should work, but rather how in fact it does.

THE ULTIMATE ASSESSMENT MECHANISM

Despite the vast amount of thought and attention being devoted to realizing a system of rational management of technology assessment by a democratic political process, this goal will not be easily achieved. In fact, it may be a utopian concept, if not a contradiction in terms, since political processes are unfortunately seldom either completely rational or truly democratic, let alone both. Alternatively, the assessment process is more likely to be an emotional conflict of special interests, as gleefully predicted a few years ago by a professor of political science:

> The question, "what are the social and political implications of R&D?" is now emerging as the next great debate

> of our time . . . . Science and technology have an honest and honorable place, but it is a limited place, rationed in accordance with responsible interest group activity and a pluralistic definition of the national interest. . . . In the decade ahead, R&D programs will be a battleground of public attention, anxiety and trouble. The scientist, engineer, systems manager, and industrialist will be embroiled as never before in politics. Outsiders, reformers, and new pressure groups will play an increasing role, and the technical specialists, in outraged impotence, will sputter that if only they were left alone, they could solve every problem.[10]

The more important the issue, of course, the more intense will be the conflict and the more strident the rhetoric. Only last year we saw this prediction borne out in classic fashion by "the great ABM debate," in which the technical controversy (with Nobel Prize winners on both sides of the argument) was overwhelmed by an emotional public furor expertly programmed by what has been called the "liberal-academic complex."[11] Young, bearded members of the American Physical Society picketed the White House. Senate hearing rooms were crowded with women wearing "Stop ABM" buttons. Personals in The New Republic asked "Want to Shoot Down the ABM? Send for . . . . ," or advertised bumper stickers, "ABM is an Edsel" or "ABM is a BM." Cartoons in The New Yorker and The Washington Post lampooned the system. In the end the ABM authorization squeaked through the Senate, not on the strength of any technical arguments, but rather on political "gut feelings," considerations of party loyalty, and reservations about challenging the authority of a new President to carry out his constitutional responsibility of defending the nation. This year the outcome may well be reversed.

In short, assessment of not only secondary physical and ecological effects but of the social and political impacts of major development programs becomes a matter of public policy. Whereas the scientific and technical community will be required to expose the questions and delineate the arguments, the "go-no go" decision must be thrashed out in the forum for discussion and development of public issues, which is Congress. With its ready access to the top scientific and technical talent in the country via consulting panels and arrangements with the National Academies of Science and Engineering and to the leading authorities in all sectors via its hearings and provided with scholarly staff support by the Science Policy Research Division of its Library's Legislative Reference Service, Congress has acquired an expertise of its own on a wide range of technical subjects in which it used to defer to the superior wisdom of august committees of the Executive Branch.

Those days of obeisance to the Executive are gone forever. On minor or simple and clear-cut issues the recommendations rationally drawn in the cloistered calm of a departmental or Presidential scientific advisory board may be accepted on authority by the Legislative Branch and go unnoticed by the public. Big, important, and complex issues, however, will increasingly come under public scrutiny during the legislative process, and the impact and implications of new technology will surely be among them.

Running the gauntlet of this assessment will no doubt be painful for all new technological developments and fatal for some. The too widely voiced fear that technology is a juggernaut that has developed an irresistable momentum of its own is a myth. Just the opposite is the case: technology flourishes only in the proper political and intellectual climate, as the laments from our European friends on the other side of the "technological gap" only too plaintively attest to. As pointed out by Nieburg,

> Scientific and technical change are far from unstoppable and automatic, but are rather the result of and responsive to public policy . . . . That means more politics, troublesome, unavoidable, and good for the nation.[12]

Painful though the political process may be for technology, it is nevertheless democratic when conducted under the proper "rules of order", in which case a "pluralistic definition of the national interest" can lead to a broad consensus reflecting collective "wisdom" rather than the narrow expertise of a specialist elite, be it ever so rational.

For a democratic assessment process to function, however, it is essential that the "rules of order" be followed, i.e., that all adversary opinions receive a fair and equal hearing in an unemotional atmosphere. That this requirement is becoming increasingly difficult to satisfy constitutes a major and growing threat to technology. Again, the "great ABM debate" provides a vivid example of how rationality can only too easily be trampled underfoot. However valid, the technical arguments pro or con the system had nowhere near the impact on public opinion (and hence pressure on Senators) than did the vitriolic Herblock cartoons on the one side or the apocalyptic views of impending doom on the other. Clearly, the principal danger is not that new technology will be furtively imposed upon a blissfully ignorant public without adequate assessment, but rather that it will be overassessed to the point of harassment by hysterical (and hardly democratic) scientific Philistines, principally from the sinister side of the political spectrum.

THE ATTACKS ON SCIENCE AND TECHNOLOGY

Onslaught from the Left

Such harassment is not a conjectured possibility: it is already happening. Particularly vulnerable to irrational attack are "misuses" of science and technology for military purposes. At MIT, the technology of inertial guidance has been "assessed" by certain self-anointed inquisitors and found to be "immoral" (if not downright evil) and unworthy of future advancement under the aegis of the Institute, especially not under the continued direction of "war criminal" Stark Draper. However, simple divestment of the Instrumentation Laboratory from MIT was not permitted by these moralists; the filthy Federal dollars derived therefrom covered too much of the overhead of the greater academic institution. Rather, its work is to be redirected from inertial guidance toward more "socially constructive" purposes. Other universities which through either thoughtlessness or greed have allowed classified World War II research projects to become campus fixtures are experiencing similar withdrawal pains.

Whereas classified research is not normally appropriate to academic institutions, unclassified research along lines potentially relevant to the national defense certainly is, and the growing estrangement of some members of the university scientific community from their colleagues in government and industry who are responsible for the defense of the nation thus poses a potentially serious threat to our future security. This threat has not gone unobserved:

> Like it or not, our defense establishment is intimately interwoven with our scientific effort (or should it be stated the other way around?). It is hard to see how this tie could be unraveled without grave disservice to the country.[13]

Such warnings notwithstanding, every day's news shows that this estrangement is, if anything, accelerating. In this context it is ironic to note that the United Kingdom, having enviously assessed the benefits flowing from the U.S. "military-academic complex," has taken active steps to create one of its own; arrangements have been concluded with several universities (including Oxford) "to create an additional academic post in an existing faculty to specialize in the defence aspects of its discipline."[14]

Not only are those aspects of science and technology having military relevance under attack, moreover; but so are science and technology in general.[15,16] The deep, pervasive but hitherto unspoken fears of new technology are now being articulated with fervor and eloquence by spokesmen of "the other culture." An increasing fraction of the "developed" world's youth views science and technology not with mere disinterest but with active revulsion as having produced a "dehumanized" society, a befouled environment, and an everpresent threat of nuclear annihilation. In the universities this anti-scientific and (especially) anti-technology emotionality is not limited to the youthful students and the "New Left" but is spreading to the faculties as well, where mature professors are suffering from guilt complexes verging on self-flagellation and surrendering to the Marcusian accusation (discussed later) that science is an inhumane discipline. At the March 4, 1969 "research strike" at MIT, this theme emerged with Victor Weisskopf, the head of the physics department, quoting Winston Churchill's statement that "the Stone Age may return on the gleaming wings of Science" and one of the graduate organizers wondering "if things wouldn't be better if science and technology were stopped entirely."[17] A similar call for "technological abstinence" was voiced later last year by Murray Gell-Mann, winner of the 1969 Nobel Prize in physics.[18] At the recent Boston meeting of the American Association for the Advancement of Science (AAAS), Harvard and MIT students conducted a critique of "The Sorry State of Science" and a Columbia graduate proclaimed: "Revolution is the only solution to the environmental problems of this country."[19]

Examining this worldwide "romantic rebellion" against technology, Don Price has perceived that "science and technology, far from being considered beneficent instruments of progress, are identified as the intellectual processes that are at the roots of the blind forces of oppression."[20] In the much-quoted words of Paul Goodman, "And inevitably, given the actual disasters that scientific technology has produced, superstitious respect for the wizards has become tinged with a lust to tear them limb from limb."[21] Goodman correctly distinguishes the present anti-scientific bent from the simple machine-breaking of the Luddites, victims of technological unemployment by the Industrial Revolution in England, and hence a special interest group. Nevertheless, the Luddite rebellion may be what Price is referring to when he distinguishes the anti-scientific "New Left" movement from Marxism (or "scientific socialism") as "the first international radical political movement for two or three centuries that does not have material progress as its purpose. Far from proposing to use science and technology to improve the material welfare of the poor, it rejects technological progress as a political goal. Far from calling on government to distribute the fruits of technology more equitably, it denounces big organization in government and business indiscriminately." At any rate, as recounted by the current doyen of technology assessors, it was the desecration of the environment by the filth and squalor generated by the Industrial Revolution and its "dehumanizing" effect on society which prompted the protests by "social activist" artists and writers such as John Ruskin, Thomas Carlyle, Charles Dickens, and William Morris, whose works contributed so importantly to the origins of socialism.[22] Not surprisingly, these same factors contribute to the social malaise in today's "developed" countries, and the protests against them again center in intellectual and artistic circles.

The philosophical roots of today's "romantic rebellion" of youth have been explored by Price and others.[23,24] Within their own generation the most articulate international "philosopher-hero" of the young rebels is Daniel Cohn-Bendit, who sees the student insurrection as "a revolt against the specific horrors of modern society." The university he views with contempt as a factory to

> . . . churn out the trained personnel that is so essential for bureaucratic capitalism . . . . The production of the maximum number of graduate workers in the minimum time calls for increasingly close contacts between the universities and industry, for the ever greater adaptation of education to specific industrial needs.[25]

In general, however, the student radicals tend to sit at the feet of senior "gurus," of whom the favorite is Herbert Marcuse, who preaches that "Science, by virtue of its own method and concepts, has projected and promoted a universe in which the domination of nature has remained linked to the domination of man--a link which tends to be fatal to this universe as a whole."[26] Marcuse does seem to believe that there is a way out of this dilemma, however, either by stopping technology at a given point or by redirecting it according to humanistic values--hence the term "Marxist humanism." Others sharing this hope of "humanizing" technology are Erich Fromm[27] and Paul Goodman.[28] Less optimistic is Jacques Ellul,[29] who has been described as "the rare mortal who would not hesitate to plunge himself and all mankind backward in time."[30]

Considering the viewpoint of their gurus, it is easy to see why the student radicals and their faculty sympathizers, both here and abroad, are found principally in the departments of the humanities and of theology and sociology.[20,25] Price concludes that the danger (to science and technology) from these schools of "soft" studies comes "not because they are politically powerful but, rather, because as Mr. Marcuse suggests, they may have convinced scientists themselves that science is an inhumane discipline."[31] The anguished orgies of expiation at the MIT "research strike" and the AAAS meeting are thus perfect examples of the trend discerned by Marcuse in which dehumanized science "feels the need for redemption by coming to terms with the humanities."[31] Such, then, is the philosophical basis on which the "new humanism" sweeping this country and others is causing many of today's brightest and most idealistic students to eschew the physical sciences and engineering for the humanities and social or life sciences as more closely identified with their desires of building a better society.[32,33]

Ambush by the Center

> I see storm signals flying. The pressures of rising expectations affect scientists and engineers as well as ghetto dwellers. Our people's commitment to science is an act of faith that carries with it the dangers of disappointed expectations.[34]

Compounding the difficulties faced by science and technology, this "thunder on the left" is paralleled not yet so much by a "reaction on the right" as by an unintentional ambush by centrists threatening basic research in particular--a Baconian demand for "science assessment," which nicely complements the attack on technology described above. A growing public disenchantment with science and skepticism about its general relevance to the public interest was first detected by sensitive political antennae:

> A great deal of basic research has been done. . . . I think the time has now come to zero in on the targets by trying to get our knowledge fully applied . . . . I am going to show an interest in the results.[35]

> Congress wants scientific research that gets results.[36]

With the growing public demand for visible payoff from R&D programs supported by its tax dollars thus given Presidential affirmation, the "name of the game" of justifying Federal research support quickly became "<u>relevance</u>." This criterion now applies to "basic" research as well as to work with more easily identified objectives, thus augmenting if not dominating the classic "internal" criteria previously applied by the elite "new priesthood" prior to its defrocking.[37] The power of the new criterion was early manifested in the Congressional rewriting of the charter of the National Science Foundation to include sponsorship of applied research. (The trauma of the NSF in facing the application of cost-benefit considerations to the evaluation of its research proposals is easy to appreciate.)

In keeping with growing public awareness of mundane societal problems, the relevance criterion has become more specific and now is interpreted as meaning "relevant to the problems of society," and the demand for a "reordering of national priorities" has become more strident. So far, the most publicized victim of this "reordering" has been the NASA manned space program. The triumph of Apollo 11 in winning the Moon Race was glorious, but subsequent missions are suffering from declining Nielsen ratings. As reflected in the President's fiscal year 1971 budget request, Man-in-Space is losing its popular "pizazz" to poverty and pollution. A more crippling if less spectacular blow, however, was sustained by academic scientists previously supported by NASA. University research sponsored by the Office of Space Science and Applications will drop almost 17 percent, and the Sustaining University Program will be terminated.[38,39] Declines in university research supported by the AEC and HEW will also occur. That the National Science Foundation budget will increase somewhat[40] is "largely illusory when measured against the new responsibilities NSF is inheriting as castoffs from other agencies."[41] Condemning such "dumping," science's most understanding friend on Capitol Hill traced the origin of the problem to a delay until 1950 in forming the National Science Foundation following World War II:

> In that interim the mission agencies moved in with big money, and NSF never caught up. Now some of these same agencies are looking towards NSF and saying, "Here, catch!" as they have second thoughts about the nature and extent of their responsibilities for academic research. In this time of financial constraint, NSF finds real obstacles in getting funds not for new research but to take care of orphans of the fiscal storms.
>
> If ever there was a matter that requires national policy, this is it.[42]

The greatest "dumper" of research programs, ironically enough, is motivated not by self-interest but by deliberate legislative action. By far the most serious threat to university research programs is posed by Section 203 of the fiscal year 1970 military procurement authorization bill, which provides that, "None of the funds authorized to be appropriated by this Act may be used to carry out any research project or study unless such project or study has a direct and apparent relationship to a specific military function." Introduced by the Senate Majority Leader, this statement constitutes an unequivocal criterion of "relevance" by which Department of Defense (DOD) sponsored research will be judged. Clearly, Senator Mansfield does not share the viewpoint of the estranged academics concerning the military sponsorship of university research:

> This is not to say no university research should be funded by the military. But those who seek and receive such funds should realize that they are not enjoying a form of scientific largesse from a grand patron, but rather are joining their talents and interests with the needs of the military for the defense of this country, which in itself is honorable and necessary.[43]

Senator Mansfield further emphasized his point by quoting the agreement of DOD's John S. Foster:

> Research is done to provide a technological base, the knowledge and trained people, and the weapons needed for national security. No one in DOD does research just for the sake of doing research.

and citing the historical conclusion of Derek J. deSolla Price:

> From the time of Archimedes and Leonardo onwards, they [the scientists] have been able to demonstrate conclusively to any government that maximum support of every need of scientific work was essential for the military and economic security of the state. Unfortunately, the demands of scientists now begin to exceed the possibilities of support of the largest and most developed countries. We, therefore, begin to have a problem of "overdeveloped countries" where one must somehow learn to say no to at least some of the reasonable demands of the scientific community.

He concluded his speech with a declaration of serious intent to enforce the public law:

> During the coming year we will look to the Comptroller General and his auditors to give close attention to section 203. I, for one, will expect them to establish that

research administrators of the Department of Defense provide a written determination of the need for and relevance of each research project to military science and technology. Each DOD administrator should have to establish the relevance of each project at the time the decision is made to fund it, not after a Congressional inquiry generates a demand for rationalization. Furthermore, I will ask the Comptroller General to report what changes in research procedures he has observed that carry out section 203.

Congress, in writing section 203, has stated an important national policy; it must be vigorously adhered to by those responsible for the expenditures of these funds.

The stated intent of section 203 is to correct the situation whereby a substantial fraction of Government support for university research has, since the founding of the Office of Naval Research following World War II, been provided by the Department of Defense. Noble though this intent may be, it poses a dilemma for the university researcher: if his research passes the test of relevance to "a specific military function," it is sure to put him on a collision course with the radical faction among the students and his academic colleagues. For the ideological purists who are clamoring for scientists to boycott all DOD-supported research (or "turn themselves off" in Bernard Feld's hip jargon) there is no problem: they simply ignore the defense research requirement as immoral, obscene, or criminal. One simply cannot imagine a supplicant Noam Chomsky, for example, justifying support of his research on linguistics by pointing out that the fruits of his labor are applied by the CIA and the defense intelligence community for computer translation of the Soviet technical literature. For the moderate majority, however, the horns of the dilemma will be painful, and it has already been noted that "there is a rift in the scientific community on the need for military [research] spending that is not soon likely to close."[44]

The Defense Department's official line is to dismiss this problem as overexaggerated, but some veterans of student confrontations do not agree. Consider, for example, the prediction of Dr. Paul Saltman, Provost at Revelle College of the University of California at San Diego (which, not incidentally, is Herbert Marcuse's home base):

> What happened at Stanford last spring is the model of what is going to happen at UC this year. There is going to be such a hue and cry for the stopping of military research on university campuses that the students--and probably

> the faculty--will bring Berkeley to a screeching halt until it disperses all of its holdings in Livermore and Los Alamos.
>
> Students at Caltech--it may take them a little longer--are going to be investigating every grant that any faculty member ever had and any consulting that he does for any agency of the Federal Government or private industry that has in any way, shape, or form a relationship to the military-industrial complex--whatever that may be. That's going to be the real low point. That's when the real barricades are going to have to be built.[45]

If this vision appears pessimistic, even more ominous omens have been noted. At the Boston meeting of the AAAS, a specialist in social medicine urged U.S. scientists to break security and inform the United Nations about secret work on chemical and biological warfare.[46] Noam Chomsky has offered the grim prediction on BBC television that the conscience of some U.S. scientists may lead them to actions morally justified "for the safety of mankind" but illegal before the law, as in the cases of Alan Nunn May and Klaus Fuchs.[47]

Reaction on the Right?

As noted earlier, no specific instance of repressive reaction against scientific inquiry has yet been noted coming from the political right, but some academic scientists as exemplified above are surely courting one and certain indicators of a popular reaction are already discernible. The organized disruption of DOD-sponsored research at universities may exhort the faithful of the New Left, but is viewed with dismay by the "silent majority," as manifested in demands for punitive legislation. The calls for actions ranging from research boycott to quasitreason have, to say the least, a limited appeal; unlike Lysistrata and her _felenades,_ scientists of this pursuasion do not have a corner on the desired commodity. More damaging to the public image of science is the participation of some scientists in radical violence. The recent acquittal of John Froines (Ph. D. Yale, Member American Chemical Society, AAAS, The Royal Institution, Sigma Xi, and the "Chicago Seven") on charges of inciting a riot, teaching the making and uses of incendiary devices ("defensive chemistry"), and obstructing fire and police forces in the performance of their duties is interpreted by the general public not as denoting his innocence of these actions but as resulting from the expert invocation of fine legal niceties. How many more scientists, it must wonder, regard as an important objective (to quote Froines' own words) "the subversion of the democratic process"?[48]

Possibly the most damaging blow to the image of science, however, is the revelation by their own words that its greatest men are not really endowed with superior wisdom but can be every bit as irrational and ignorant outside their specialties as the unpresumptuous populace. A common weakness of scientists which limits their usefulness in government is a peculiar proclivity to pronounce profound judgments on subjects far beyond the scope of their professional expertise: "All too often the scientist fails to recognize that he has gone beyond the boundaries of what can be proved by research and is speaking _ex cathedra_ on matters on which his own judgment is just as personal, and perhaps nearly as prejudiced, as any layman's."[49] Perhaps the most publicized recent example of this scientific syndrome was provided by last year's "March 4 Movement" at MIT, where, in a passionate speech, biologist and Nobel laureate George Wald of Harvard declared some overblown but otherwise accepted political rhetoric by the President _Pro Tempore_ of the U.S. Senate to be "criminally insane."[50] While evoking "prolonged applause" at the time, such hyperbole also provoked questions in the minds of less emotionally aroused individuals concerning Dr. Wald's competence in other fields besides psychiatry and criminal law, specifically politics (foreign and domestic) and national defense. Although he wowed his academic audience and the liberal press, Wald's pronouncement contributed just one more shadow of doubt to a growing public disillusionment as to the omniscience of scientists, even Nobel laureates. In the eyes of a George Wallace, he would be dismissed as a "pointy-headed intellectual."[51]

Thus it can be roughly said that while the political left is becoming disenchanted with the beneficence of technology, the center and right are becoming disillusioned with the relevance of science to the public interest and the infallible wisdom of scientists, respectively. Augmenting these growing reservations regarding science and technology is the culmination of a general fear held by the entire U.S. public--if not the world--fear of the consequences of environmental pollution, as discussed later.

BELATED RECOGNITION OF DANGER SIGNALS

In view of the fact that the attack against science and technology had been visibly building since 1963, it is amazing that the storm warnings could go so long unrecognized by the scientific community. Whereas the technologists are sensitive to changes in the political climate (particularly so if engaged in defense technology), the scientists only recently have awakened to the fact that times have changed since the halcyon era of affluence and influence described by Don

Price: "Yet the plain fact is that science has become the major Establishment in the American political system: the only set of institutions for which tax funds are appropriate almost on faith, and under concordats which protect the autonomy, if not the cloistered calm, of the laboratory."[52] The end of that era was marked when, on September 11, 1963, the House of Representatives by unanimous vote authorized the establishment of Select Committee to investigate the causes of the unprecedented size of the Federal R&D budget. The first progress report of this Committee contained an unmistakable warning:

> The Federal Government's marriage to research and development has been marked by an amazingly long and luxurious honeymoon. This is due mainly to the exhilarating nuclear age atmosphere in which the union was finally fused and unsparingly nurtured.
>
> Noting the recently slackening annual increase of Federal funds for research and development activities, some say the honeymoon is over.[53]

Clearly, 1963 was "the winter of Congressional discontent with science,"[54] following which the aforementioned "marriage" rapidly became one of necessity rather than amour. Despite the unmistakable character of this trend, the realization that the "good old days" are gone forever, and that science is indeed "on the street with the tough guys"[55] seems to have come as a shock to most of the scientific community, as manifested by the "vale of tears" atmosphere prevailing in scientific conferences during the last year or so.

Unlike the case of the technological community, which had long sensed the coming onslaught of the "liberal-academic complex", only last year did it fully dawn on the scientists that not only were times getting tough but that they, too, were actually under attack, a situation which they find "incredible."[56] In the words of a new university president,

> The basic values of our educational institutions and our society are now being contested. Our universities are under scrutiny and under attack. Rational thought is under scrutiny and under attack. And science and technology are under scrutiny and under attack. These institutions and values, until very recently, were the most respected of all.[57]

Taking note of the fully developed attack on science and technology Harvey Brooks has observed its obvious consequences:

> This hostility has spread in our time from a small literary elite to a wide section of the educated public, especially some of our most highly educated youth. Not only is less importance attached to material progress as a condition of a viable society, but also there is an almost obsessive preoccupation with the bad side effects of technology while taking its benefits for granted. If we assume even a moderate degree of persistence in this present mood, the outlook for technological change appears rather dim.[58]

In even the short run it can easily produce what Lee DuBridge terms "technology arrestment."[59] Noting that "enthusiasm for technological progress is deeply implanted in the Soviet Union, and continues to flourish there," Arthur Kantrowitz views with alarm still another consequence:

> I am convinced that only a few more years of the present vigorous counterattack on technology will result in a technological gap between the U.S. and the Soviet Union which will be reflected in our economic and our strategic posture with catastrophic consequences . . . . We must not allow those who would escape this challenge to dominate the thinking of young people.[60]

In the longer run, even basic science in the U.S. could well wither in the arid climate of student disinterest if not hostility, public indifference or disenchantment, and Congressional skepticism. During the year elapsed since his emotional March 4, 1969 warning that "the Stone Age may return on the gleaming wings of Science," the gravity of science's true predicament and the real danger of its withering have become evident even to Victor Weisskopf, who now sounds the alarum that "This ominous trend must be combated."[61]

A TIME FOR REDEMPTION?

> This is the way the world ends
> This is the way the world ends
> This is the way the world ends
> Not with a bang but a whimper.[62]

In his remarkable "underground" monograph of 1967, the "father of the Soviet H-bomb" rated pollution of the environment on a par with thermonuclear war as a threat to the continued existence of mankind.[63] Now, only a few years later, it might appear that the environmental threat is the greater, if less cataclysmic, because of its insidious nature. During the interim, public awareness of the fragility of Spaceship Earth's life support system has grown rapidly, and the voices of Rachel Carson and Barry Commoner have been joined by a broad chorus including Pogo and the President of the United States. Ecology is the new "OK word" on campus and in legislatures alike, and vivid doomsday scenarios dramatize the issue.[64] U.S. Senators vie for environmental championship, national magazines devote special issues to the subject, and improved environmental quality is acclaimed as a national goal for the next decade. In the tempered judgment of a bona fide ecologist with impeccable academic credentials,

> Concern for the environment has reached hysteric levels and become a major public issue in this country. Perhaps the hysteria is needed. It is the wrath of the people and not the evidence of environmental deterioration that is causing national priorities to be changed.[65]

Possibly to the relief of the Nixon administration, the environment is eclipsing Vietnam as a focus of student passion, and nationwide "teach-ins" are scheduled for "Earth Day," April 22, 1970. The growing realization that the basic root of the problem is exponential population growth ("popullution") has not yet dawned on the general public, however; and the search for surrogate scapegoats proceeds apace. The traditional bete noir of conservationists has been industry, particularly the petroleum and chemical process, pulp and paper, mining and metallurgical, and electric power industries. That in the past this unsavory reputation has been well-earned is undeniable, as the weirdly beautiful if grisly color photographs in recent issues of *Time*, *Newsweek*, and *Fortune* magazines plainly attest. More recently, a new villain has been identified in the automobile industry, whose products stand guilty of poisoning the air, clogging the city, and desecrating the landscape. What generally goes unrecognized (or at least unemphasized), however, is that the culprit is not technology per se but persistence in the application of obsolescent (if not archaic) technology for economic reasons, and failure to apply new or existing technology for the processing of waste products and recycling of useful products, once used. What is needed is not less but more and better technology, thoughtfully applied. Now that the "wrath of the people" is being expressed, all industries are moving rapidly if belatedly to modernize their technologies, often at considerable cost.

That this cost must be passed on to the consumer in the form of increased price of product, however, is regarded by too many people as "unscrupulous business practice" rather than recognized as the inevitable workings of economics. There somehow persists an article of faith that a clean environment should not only be an inalienable right of everyone,[66] which ideally it should, but should be free as well. Public awareness that there is an "external" cost associated with environmental quality and that "industry" alone cannot pick up the tab is not well enough developed. An illuminating example of this public attitude of "let the other guy pay" has been cited by Louis Roddis:[67]

> Around 1960, California laws first required the installation of closed crankcase ventilation systems on new cars. This step was intended to eliminate about 25 per cent of unburned hydrocarbon emissions from the vehicle: an end both desirable and feasible. The price of the addition was not obvious in the total new car price, and so the public rather felt that "industry" was paying for this benefit.
>
> Then, some three years later, the California Motor Vehicle Pollution Control Board began requiring the installation of similar devices, costing $10 to $20, in all existing cars.

The result was that "all hell broke loose":

> The first time the California "public" had to reach in its wallet and come up with the price of admission to cleaner auto exhausts was when crankcase controls became mandatory on used cars in metropolitan counties in 1964 and early 1965.
>
> A pink card explaining the law was mailed to 10,000,000 California motorists. The card evoked almost universal rejection, and was then disavowed by practically every spokesman for State Government. The press decried the "infamous red card." An emergency law was rushed through the State Legislature, placing an immediate moratorium on crankcase installation for used cars. Another bill, which would have abolished the MVPCB itself, went as far as passage in the State Senate.
>
> Three counties, using county option, reversed a previous position, and voted themselves out of the program for used vehicles. More threatened to do so.[68]
>
> All this in a state where there is widespread understanding of the part the automobile plays in atmospheric pollution. What was not widely understood--or at least not

> accepted--was the notion that in pollution control you pay for what you get, either directly or indirectly.[67]

Other examples of public failure to recognize Pogo's observation that "We have met the enemy and they are us" are familiar. The siting of new power plants (especially nuclear plants) is becoming almost impossible because of protests by a local populace which, on the other hand, increasingly demands the comforts of air conditioning and electric household slaves. Municipal waste disposal systems are overwhelmed by the throw-away containers of an "effluent society." What is required to cope with the environmental problem is public recognition that it will have to change its life style and control its own numbers. Acting in the role of a responsible spokesman for industry in assessing its own technology, Roddis also made the following cogent points:

> Industry is no monolith. Yet increasingly today, particularly among young people, those of us in industry face being stereotyped and dehumanized in the rhetoric of environment. It is as if "industry" has life apart from society; and when industry speaks on the subject it is in defense of pollution, and as an adversary of beauty, health, and even life. Industry has no life apart from society. To the extent I may speak for industry, I trust we will not contribute to what I think is a danger in society's confrontation of its environmental responsibility: faddism.
>
> Many have donned concern for the environment like a hula-hoop. Awareness threatens to become frenzy. I applaud the rise in public consciousness of this vital issue. Scarcely a public official lives who has not embraced it. But let us be sure it doesn't die in that embrace. Let us be on guard that when the media's attention flags, when dire warnings are repeated so often they echo revival-tent descriptions of hell--with as little effect on life style--that, in short, we don't get bored.
>
> Restoration of the environment will not occur in a crescendo of attention on "Earth Day," April 22. Nor will action, however concerted, bring it in a quantum jump on any other day we can foresee. It will happen bit by bit, reversing the gradual process of accretion by which the present state of deterioration came about. April 22 will have real meaning only if it begins transforming our laws and their enforcement, our system of taxes and incentives, and our whole educational process.

> Environmental concern must become an integral part of the ongoing functions of production and consumption. For so long as this concern remains only an adjunct to business as usual, whether by producers or consumers, industry or government, we will not get the result we seek.[67]

It is ironic that, well in advance of the current furor, the most articulate early warning by a Government official about the consequences of man's rape of the biosphere and call for a humanistic technology came from a military officer. Observing that (pure, not applied) "science is the antithesis of 'humanism,'" Hyman Rickover went on to state the following:

> Science, being pure thought, harms no one; therefore it need not be humanistic. But technology is action, and often potentially dangerous. Unless it is made to adapt itself to human interests, needs, values, and principles, more harm will be done than good. Never before, in all his long life on earth, has man possessed such enormous power to injure his human fellows and his society as has been put into his hands by modern technology.
>
> This is why it is important to maintain a humanistic attitude toward technology; to recognize clearly that, since it is a product of human effort, technology can have no legitimate purpose but to serve man--man in general, not merely some men; future generations, not merely those who currently wish to gain advantage for themselves; man in the totality of his humanity, encompassing all his manifold interests and needs, not merely some one particular concern of his. Humanistically viewed, technology is not an end in itself but a means to an end, the end being determined by man himself in accordance with the laws prevailing in his society.[69]

Rickover's warning did not go unheeded, even in industry. In the one case on which he can speak with first-hand personal authority, two years ago the present author emphasized to his corporate fathers that a primary assessment criterion for justifying Federal support of new technology would increasingly be its compatibility with the human environment.[70] In the case of the aircraft industry the workings of this criterion are now manifested in Federal Aviation Authority requirements that the new generation of subsonic jet transports be powered by engines that are smoke-free and less noisy than those in current use. Roddis' words confirm that in many cases this criterion is also being applied by other sectors of private industry. This belated trend does not mean so much that technology is "starting to develop a heart"[71] as that its users are "starting to develop a head," namely, a longer view of their enlightened self-interest and public responsibilities alike. What seems to be happening is that, as perceived by Marcuse, science and technology are indeed feeling the need for "redemption" by coming to terms with humanity. Examples abound. The theme of Caltech's development program, for example, is "Science for Mankind," and its new president declares that "We must resolutely defend the rational process, science and technology, <u>and</u> the human values from the know-nothings of our day."[72] Although distorted by its anti-military hysteria, MIT's "March 4 Movement" contained the same theme. A symposium on "Science and the Future of Man" at the Boston AAAS Meeting called for applications of science to society's problems,[73] in a sense extending Rene Dubos' earlier proposal that "we need to develop a science of humanity and of prediction of consequences--to supplement the science of things and of effects in the here and now."[74] Paul Goodman's passionate contention that "technology <u>is</u> a branch of moral philosophy"[75] is echoed by Paul Johnson:

> We are beginning to recognize, somewhat fearfully, that human society cannot absorb major scientific and technical developments without qualitative changes in its structure and attitudes. We cannot just use machines to make our lives the same but better; they also make life different--often, in some respects, worse . . . . What we really need is a new moral philosophy, capable of public application.[76]

Increasingly it is being realized that the power of science and technology is approaching the point where the future can be designed:

> The present challenge to the public intelligence--that of rapidly advancing science and technology--derives from the dawning realization that man is on the point of being able to do anything he wants . . . . But if the answer to the question "What can we do?" is "Anything," then the accent shifts to the questions, "What should we do?"; "What, in the end, do we really want?" We take up again, in other words, the task of the ancient Greeks: How to be wise. I might add that it is an infinitely more difficult task now than it was for the Greeks, because we have infinitely more power than the Greeks had.[77]

Since wisdom is a matter of values, the very power of our technology is posing increasingly difficult value judgments, responsibility for which cannot be arrogated by a scientific or technical elite. This power also has the effect of "adding new faces to the dice" and permitting changes in society as well as nature: "The development and adopting of new technologies make for changes in social organization and values by virtue of creating new possibilities for human action and thus altering the mix of options available to men."[78] Thus the stage is set for the thrashing out the "pluralistic definition of the national interest" foreseen by Nieburg[79] by a lusty political process of which an essential element must be vigorous adversary proceedings.[80] The formal establishment of a special agency, institution, or mechanism for such proceedings may be a long time in coming. In the meantime the issues will continue to be fought out not only in closed committees or advisory boards but in the hearing rooms and on the floors of Congress, in the press and publications media (including the "underground" press), probably in the telecommunications media, and certainly upon the campus. Everybody will be in the act in what Harvey Wheeler predicts will be "a new kind of participational democracy," a new kind of politics "especially designed to cope with science and technology, a politics based on what Teilhard de Chardin called the 'noosphere' instead of our accustomed politics based on real estate."[81] A major concern will be to avoid "technology assessment on the str streets," but such a possibility can by no means be discounted. In any event, the public fray is certain to be spirited, if not painful, and major new technology programs will have to measure up under a trial by fire.

CONCLUSION

This paper has argued that important issues of science, technology and public policy cannot be settled by the "experts" alone, but will require a pluralistic definition of the public interest by open adversary proceedings. Considering the attacks to which science and technology are now being subjected, the danger is not that new technology will receive inadequate assessment for possibly deleterious secondary effects, but that harassment by an overemotional political process may prevent its coming to fruition. Such a risk must be run, however, to assure that the new technology will meet its legitimate purpose of serving the public interest.

NOTES

1. Genesis 11.

2. U.S. House, Committee on Science and Astronautics, Subcommittee on Science, Research, and Development, Technology Assessment Seminar, 90th Cong., 1st sess. (September 21 and 22, 1967).

3. Program of Policy Studies in Science and Technology, The George Washington University, *Technology Assessment--The Proceedings of a Seminar Series* (July, 1969) (The Proceedings, with slight revisions, are included in this volume as Chapters 2-5).

4. U.S. House, Committee on Science and Technology Report of the National Academy of Sciences, *Technology: Processes of Assessment and Choice* (July, 1969).

5. U.S. House, Committee on Science and Astronautics, Report of the National Academy of Engineering, Committee on Public Engineering Policy (July, 1969).

6. U.S. House, Committee on Science and Astronautics, Subcommittee on Science, Research, and Development Hearings, 91st Cong., 1st sess. (November-December, 1969).

7. "Proceedings of the Engineering Foundation Research Conference on Technology Assessment," Proctor Academy, Andover, N.H., (August 4-8, 1969), pp. 359-501.

8. "Technology Assessment and Human Possibilities," General Symposium at the American Association for the Advancement of Science 136th Meeting, Boston, Mass., (December 26-27, 1969).

9. "Science, Technology, and the Quality of Life," Third American Institute of Aeronautics and Astronautics President's Forum, *Aeronautics & Astronautics*, VIII, 2 (February, 1970), 28-54.

10. H. L. Nieburg, "R&D Program Implementation," in *Proceedings of the Symposium on National R&D in the 1970's* (Washington, D.C.: National Security Industrial Association (1968), p. 203.

11. R. Harwood and L. Stern, "Liberal-Academic Complex Disowns its Military-Industrial Offspring," *The Washington Post*, June 4, 1969, p. A23.

12. Nieburg, *op. cit.*, p. 210.

13. P. P. McCurdy, "A Noble but Questionable Crusade," *Chemical & Engineering News*, XLVII, 9 (March 3, 1969), 5.

14. T. Dalyell, "Westminster Scene," New Scientist, XLIV (October 23, 1969), 191.

15. Arthur Kantrowitz, "The Test: Meeting The Challenge of New Technology," Bulletin of the Atomic Scientists, XXV, 9 (November, 1969), 21.

16. A. M. Weinberg, "In Defense of Science," Science, CLXVII (January 9, 1970), 141-49.

17. B. Nelson, "Mit's March 4: Scientists Discuss Renouncing Military Research," Science, CLXIII (March 14, 1969), 1175-78.

18. "Technological Abstinence Advocated," Science, CLXVI (November 7, 1969), 723.

19. James K. Glassman, "American Association for the Advancement of Science Boston Meeting: Dissenters Find a Forum," Science, CLXVII (January 2, 1970), 36-38.

20. Don K. Price, "Purists and Politicians," Science, CLXIII (January 3, 1969), 25-31.

21. P. Goodman, Utopian Essays and Practical Proposals (New York: Random House, 1964), p. 41.

22. Lewis Mumford, Technics and Civilization (New York: Harcourt, Brace and Company, 1934), Chapter IV, "The Paleotechnic Phase," pp. 151-211.

23. Philip L. Bereano, "The Scientific Community and the Crisis of Belief," American Scientist, LVII, 4 (Winter, 1969), 484-501.

24. Manfred Stanley, "The Technicist Projection: A Study of the Social Theory in Moral Rhetoric," in Freedom and Tyranny in a Technological Society, Jack D. Douglas, ed. (New York: Random House, to be published).

25. S. M. Lipset, "The Political Thrust Motivating Campus Turmoil," Saturday Review, LII (March 1, 1969), 23 A review of Daniel Cohn-Bendit and Gabriel Cohn-Bendit, Obsolete Communism: The Left Wing Alternative (New York: McGraw-Hill Book Co., 1968)

26. Herbert Marcuse, One-Dimensional Man (Boston: Beacon Press, 1964), p. 166.

27. Erich Fromm, The Revolution of Hope (New York: Bantam Books, 1968), p. 100.

28. Paul Goodman, "Can Technology be Humane ?," The New York Review of Books (November 20, 1969), 27-34.

29.Jacques Ellul, The Technological Society (New York: Alfred A. Knopf, 1964).

30. A. Rupert Hall, Scientific American, CCXII, 2 (February, 1965), 125, review of Ellul op. cit.

31. Price, op. cit., p. 26.

32. J. Walsh, "Dainton Report: British Youth Swings--Away from Science," Science, CLIX (March 15, 1968), 1214-15.

33. "Science Under Fire: Behind the Growing Concern," U.S. News & World Report (February 9, 1970), 30-31.

34. Emilio Q. Daddario, "Needs for a National Policy, XXII" Physics Today (October, 1969), 33.

35. Lyndon B. Johnson, White House Press Release, quoted in Science, CLIII (July 8, 1966), p. 150.

36. Joe L. Evins, Congressional Record, C1249-50 (February 27, 1968).

37. Alvin M. Weinberg, "Criteria for Scientific Choice," Minerva, I, 2 (Winter, 1963), p. 159-71.

38. Nancy Grochow, "NASA: Further Cuts in University Support Spending," Science, CLXVII (February 20, 1970), 1107.

39. "NASA Lowers Funds," Chemical & Engineering News (February 16, 1970), 14.

40. "R&D Budgets: More Belt Tightening," ibid., pp. 32-40.

41. Charles A. Mosher, quoted in "Federal R&D Budget Cuts Squeeze Chemists," Chemical & Engineering News (February 23, 1970), 37.

42. Daddario, op. cit., p. 35

43. Senator "Mike" Mansfield, "Department of Defense Research and Development," Congressional Record--Senate (November 6, 1969), S 13899-900.

44. "A Clash With Science," Chemical & Engineering News (September 22, 1969), 17.

45. Paul Saltman, "Who's In Charge Here?," Engineering and Science (January, 1970), 39.

46. Victor W. Sidel, "CBW as an Ethical Challenge," New Scientist (January 15, 1970), 124.

47. "For The Safety of Man," New Scientist (January 8, 1970), 43.

48. "The John Froines Affair," Chemical & Engineering News (October 13, 1969), 15.

49. Don K. Price, Government and Science (New York: New York University Press, 1954), p. 133.

50. George Wald, "A Generation Unsure That It Has a Future," The Washington Post (March 30, 1969), B3.

51. According to Irving Kristol, "American Intellectuals and Foreign Policy," Foreign Affairs, XLV (July, 1967), 594: "An intellectual may be defined as a man who speaks with general authority about a subject on which he has no particular competence."

52. Don K. Price, "The Scientific Establishment," in Scientists and National Policy Making, R. Gilpin and C. Wright, eds. New York: Columbia University Press (1964), p. 20.

53. U.S. House, Select Committee on Governmental Research, First Progress Report, Federal Research and Development Programs, 88th Cong., 2nd sess. (Washington, D.C.: U.S. Government Printing Office, 1964).

54. Kenneth Kofmehl, "Science and Congress," Annual Meeting of the Midwest Conference of Political Scientists, Madison, Wisconsin (April 30-May 2, 1964).

55. D. S. Greenberg, "Science--On the Street With The Tough Guys," Astronautics & Aeronautics, V (March, 1967), 17-18.

56. Weinberg, op. cit., p. 141.

57. Harold Brown, Inaugural Address, "The President's Report 1968/1969," *Bulletin of the California Institute of Technology,* LXXVIII, 4 (December, 1969), 3.

58. Harvey Brooks, quoted in "Science Under Fire: Behind the Growing Concern," *U.S. News & World Report* (February 9, 1970), p. 31.

59. Statement of L. A. DuBridge, Ref. 6, p. 221.

60. Kantrowitz, *op. cit*., p. 21.

61. Victor Weisskopf, "The Need for Basic Research," *Science,* CLXVII (February 15, 1970), 191.

62. T. S. Eliot, "The Hollow Men" (1925).

63. Andrei D. Sakharov, *Progress, Coexistence & Intellectual Freedom* (New York: W. W. Norton & Co., 1968).

64. Paul Ehrlich, "Eco-Catastrophe!," *Ramparts,* VIII (September, 1969), 24-28.

65. Frederick E. Smith, "Scientific Problems and Progress in Solving the Environmental Crisis," presented at the joint International Biological Program/Public Affairs Council conference on Environment: The Quest for Quality, Washington, D.C., February 18-20, 1970.

66. Statement of Charles Maechling, Jr., "The Emergent Right to a Decent Environment," Ref. 6, pp. 342-48.

67. Louis H. Roddis, Jr., "The Responsibility of Industry Toward the Environment," Ref. 65.

68. D. A. Jensen, "The Public's Role in the Automotive Exhaust Emissions Program," Society of Automotive Engineers Paper 660103 (January, 1966).

69. H. G. Rickover, "A Humanistic Technology," The Granada Guildhall Lecture of the British Association for the Advancement of Science, London, October 27, 1965 (reprinted in Ref. 6, pp. 301-09).

70. Leon Green, Jr., "Changing Criteria for Federal Support of R&D," Lockheed Aircraft Corporation Long-Range Planning Meeting, April 2, 1968.

71. Richard A. Carpenter, "Technology Assessment--A New Responsibility," Ref. 6, pp. 361-68.

72. Harold Brown, op. cit., p. 4.

73. "Scientists Seek New Goals for Science," Chemical & Engineering News (January 12, 1970), 32-33.

74. Rene Dubos, "The Unbelievable Future," presented to the Sesquicentennial Celebration of the New York Academy of Sciences (December 6, 1967).

75. Paul Goodman, "R&D for the Socio-Economic Environment of the 1970's," Proceedings of the Symposium on National R&D for the 1970's (Washington, D.C.: National Security Industrial Association, 1968), p. 190.

76. Paul Johnson, "A Morality for Dynamic Society," New Scientist (December 4, 1969), 506-07.

77. Emmanuel G. Mesthene, "Technical Progress and Social Development," in Proceedings of the 19th National Conference on the Administration of Research (Denver: Denver Research Institute, 1966), pp. 8-14.

78. ________, "How Technology Will Shape the Future, Science, XLXI (July 12, 1968), 135-43.

79. Nieburg, op. cit.,

80. Harold P. Green, Chapter 4

81. Harvey Wheeler, "The Politics of Ecology," Saturday Review LIII (March 7, 1970), 51.

DISCUSSION OF TECHNOLOGY ASSESSMENT OR TECHNOLOGY HARASSMENT

Leon Green opened the discussion by emphasizing that his paper was a summary of how he sees the assessment process working at the present time. He felt that such a descriptive approach was necessary because "it may be a long time before we get any formal mechanism for assessment established."

Mr. Green stressed his conviction that "technology is not an irresistible juggernaut . . . that cannot be stopped." Technological changes, far from being unstoppable and automatic, are the result of public policy decisions which can dictate the extent of the use of technology in our society. Mr. Green was somewhat ambivalent about the effects of "a pluralistic definition of the public interest," he expressed a fear that the process of obtaining such a definition "tends to get out of hand" and becomes the antithesis of a democratic process. His major concern was that the assessment process must be a rational one and "it's pretty difficult to have a rational process when you have all sorts of emotional forces" involved. He felt that if, as some have held, technology "has created a hell of a mess out of the world" by dehumanizing the society and befouling the environment, then "the answer is not to stop but to press on and do it in a sensible manner."

Mr. Green used two general examples of public outcry against technology to illustrate his points. One was the increasing protest against some military uses of technology. These protests arise primarily from what Mr. Green referred to as the "sinister side of the political spectrum." While granting that "there are legitimate causes for concern" over the way technology is used for military purposes, he expressed his conviction that "it will not get out of hand" because "the technology used by the government is. . .dictated by the citizens." One result of the protests, such as those at MIT, is a threat to basic research in the universities, since, as has been often noted, the Defense Department funds much of the basic research pursued on university campuses.

The other example of public involvement in technology assessment was the recent upsurge of interest in the environment. This interest "has reached hysteria." Perhaps this hysteria is needed for "it is the wrath of the people and not the evidence of environmental deterioration that is causing national priorities to be changed." In its concern, however, the "general public is searching for scapegoats" and industry, the "traditional bete noir, is taking most of the flak these days." Sometimes the criticism of the role of industry is warranted, but often it tends to obscure what Mr. Green feels is the root of the environmental problem: population growth. In Mr. Green's view industry will react to the public outcry and through the use of modern technology will alleviate some of the pressing environmental problems which face us.

Mr. Green concluded his introduction to the discussion by reading the final paragraph of his paper and emphasizing that there is a need for pluralistic assessment of the role of technology in our society, while at the same time there is a danger that technological progress will be slowed or stopped by irrational harassment.

DEFENSE AND ATTITUDES TOWARD SCIENCE & TECHNOLOGY

Some of those present felt that there was a great difference between calling for a decrease in military spending and what Mr. Green described as "technology harassment." For instance, a university researcher thought that the paper had

> . . . mixed up the opposition to the defense effort . . . with opposition to science and technology in general. There is a lot of overlap, but there is a distinction. I think that people can call for a reorientation of national priorities away from the defense effort . . . and still be calling for a larger portion of the gross national product going into R&D. . . . I would like to see more money spent on R&D and less going into defense R&D.

Mr. Green agreed that there is a difference between "this blind, irrational, 'science is no damn good' type of thing and...a legitimate beef about too much resources going into defense."

Another participant asked for a clarification of the relationship between the Mansfield amendment and technology assessment. Mr. Green responded that an important concern which stems from technology assessment is the subject of assessing the usefulness of science and technology generally and that the Mansfield amendment, in placing a relevance requirement on basic research, was requiring such an assessment. He felt that this kind of assessment of basic research was "unwarranted." But a university professor wanted to know why this kind of assessment was bad for any but a few scientists and a few universities. Mr. Green pointed out that we can never know when we have enough basic research and that "most of the work in research makes its most significant contributions largely through unpredictable applications and largely at some time in the future."

INDUSTRY IN TECHNOLOGY ASSESSMENT

Most of those present seemed to feel that the concept of technology assessment was complicated enough without being interwoven with questions of defense policy, so the discussion left the area of military uses of technology and turned to the role which industry plays in the technology assessment process.

Mr. Green reemphasized that industry's role at the present time was that of a scapegoat. As public pressure increased, he felt that industry would move rapidly to modernize its technology to avoid some of the problems which lead to this pressure. One problem that arises is that it is often not economically feasible for one company to make changes which are not made by its competitors. In order to assure industry's cooperation in cleaning up the environment, for example, "you have to have a mandate that lays it on everyone."

A university researcher felt that this was just another way of saying that "industry has no role in assessment." An industry representative felt that this was an overstatement; he noted that

> . . . industry has been assessing technology for a long while. Isn't it really the point that industry is going to have to change its way of assessing technology and to put more parameters in the assessment....

Mr. Green noted that this change is already happening.

Still another participant felt that industry did not perform technology assessments in the sense of showing a "concern for secondary and tertiary social effects" but added that "I'm not blaming industry for not doing this; I really think it's not industry's function." He contended that industry should not be put in the position of being responsible for evaluating indirect effects of technology "unless you vastly change its responsibility to its stockholders." When asked who should perform technology assessment if not industry, this participant responded that

> . . . I'm not sure I see any group within our society which has the right motivation to do it. I think that one of our major hangups is that, as a society, we haven't got institutions that worry about secondary and tertiary social effects.

When Mr. Green contended that everybody in society has to perform technology assessment, including industry, "and that's the only way it is going to work," the above-quoted participant noted wryly that "a responsibility that's held by every body never gets met."

An industry representative felt that industry's role is determined by public demand. Industry supplies what the public wants; "if they want assessments which result in fewer secondary and tertiary problems, they will get them." A university professor agreed; "industry is and will be responsive to initiatives which arise elsewhere" and we should expect no more from it.

A university researcher had a somewhat different view of the manner in which industry takes part in the assessment process. He pointed out that although the discussion had dealt with industry in general, industry "doesn't operate as a monolithic entity." Individual business firms deal with specific situations and not with broad, general notions of the public interest or social morality. He continued:

> I don't think that social morality impacts directly on industry. I think the social morality first has an impact on the legal structure . . . and through alterations in that structure changes occur in the context within which industry has to operate. The laws get changed and then industry has to function differently within them. Now it does make sense for industry to look ahead in any situation and to take account of expectations of what the laws will be in a few years. . . . But I personally think that to blame industry for the problems of technology over the last century some businessmen are more noble than others, some are less; some try to get profits in certain ways, some in others. . . . I think the key question is: What role will industry play in altering the legal structure in order to make it more responsive to social concerns? Is industry going to be resistant and recalcitrant . . .or is it going to take a longer view . . . and cooperate in the changes?

ON THE PROCESS OF TECHNOLOGY ASSESSMENT

A Government official felt that the emotional response to technology which was discussed in the paper could be avoided by performing assessments at an earlier stage in the development of a technology:

> The idea of technology assessment must be instilled in the investigator . . . far back in the industrial laboratories or the Government offices so that by the time we would get to this emotional stage, which is reprehensible and offensive, . . . we would have had the data come forward to have a rational decision on the cost and benefits of the technology.

This official felt that there was little chance that society would be swayed by the emotional reactions to the use of technology: "I predict that a technocratic society will proceed into the future with little regard for these fringe elements that are so offensive to us technocrats now, and we will achieve the great hereafter with no particular difficulty." But a university researcher was troubled by this statement and by what he felt was the tone of the paper:

> The implication is that when people show "active revulsion" toward the role of technology in causing a befouled environment, a dehumanized society, and a threat of nuclear annihilation they are reacting emotionally and that this is to be deplored. Aren't these problems grave enough to warrant active revulsion? Is there anything wrong with responding emotionally to threats to the very existence of humanity?

An industry executive was puzzled by Mr. Green's ambivalence toward a pluralistic assessment process. On the one hand, he claimed, the paper extols the virtues of a pluralistic process "yet, on the other hand, your paper argues against a pluralistic assessment coming up with any achievable results." Another participant carried this further:

> The paper was very indignant about the radicals. But if our system is such that industry cannot regulate itself by taking cognizance of the social costs of what it does for profit and that government cannot act until the situation gets almost beyond repair, then what is this kind of pluralistic, public assessment you speak so virtuously about except the actions of the radicals. If a system cannot move without people going out in the streets and screaming about it, then the radicals are necessary.

But Mr. Green felt that "going out in the streets" was not necessary and cited the example of the SST, where the assessment "is being done properly in what I would call a democratic way, which is quite different from going out into the streets."

A university researcher felt that on most issues, special interests were much more important in the assessment process than the public at large:

> On some grave public issues an aroused public will get emotional and will have an impact. But I think that on the vast majority of issues, it is the industrial leaders and other special interests working quietly with the Congress in the background that have a major impact.

Another participant was concerned with the lack of a long-term viewpoint in the decision-making bodies within our society. He felt that "as a people we tend to overevaluate the short-term and under-evaluate the long-term." This is a problem in our society, and until we overcome it any suggestions for improved technology assessment mechanisms will just be ways of "institutionalizing our short-term bias."

CHAPTER

10

PROCESSES OF TECHNOLOGY ASSESSMENT— THE NATIONAL TRANSPORTATION SAFETY BOARD

Ernest Weiss

In any assessment of technology, one must inevitably focus on real life situations to learn how assessment decisions are really made. From live decisions flow not only improvements in the technology itself but also a sharpening of the assessment process as well. The activities of the National Transportation Safety Board (NTSB), newly organized in April, 1967, when the Department of Transportation (DOT) was formed, provides some real time insight into the process of technology assessment.

HISTORY, FUNCTIONS, AND OPERATIONS OF THE BOARD AS RELATED TO TECHNOLOGY ASSESSMENT

Some history of how and why the Board was formed is provided by hearings on the bill to establish the Department of Transportation:

> . . . the Civil Aeronautics Board [CAB], the Interstate Commerce Commission [ICC], and the Coast Guard, among the other significant safety functions outlined above, conduct investigations of accidents in their respective

---

*Presented April 23, 1970. The views expressed herein are those of the author and are not intended to represent the official views or policies of the National Transportation Safety Board.

transportation fields. While the investigative procedures differ in detail, there are similarities in approach. The bill provides for a National Transportation Safety Board which is vested with responsibility for determing the cause or probable cause of transportation accidents. The Board will operate with a limited advisory staff, and will make its determinations on the basis of an independent analysis of the results of on-the-spot investigations conducted by other elements of the new Department [except for civil aviation in which it conducts most of its own investigations as well as determines the cause of all aircraft accidents.]

The Board would occupy the same position in safety matters as does its corresponding element under present law. In the determination of cause of aviation accidents, a member of the Board will preside as chairman of the accident investigative hearings assisted by a panel which he will designate as is CAB practice today. In other modes, the NTSB might assign a member to participate in accident investigative hearings but would more probably continue present practice in the Coast Guard and the ICC and act on the record of hearings conducted by the agencies.

The ultimate objective of the Board's functions is to translate the findings of accident causes into means for accident prevention. To help achieve this goal, the Board is authorized to make recommendations to the Secretary for the conduct of special safety studies pertaining to safety in transportation.[1]

Provisions of the Department of Transportation Act of 1966, describing the functions of the National Transportation Safety Board, are pertinent to this discussion:

Sec. 5(b)

. . . .

(1) Determining the cause or probable cause of transportation accidents and reporting the facts, conditions, and circumstances relating to such accidents;

. . . .

Sec. 5(d) The board is further authorized to--

(1) Make such recommendations to the Secretary or Administrators on the basis of the exercise of its functions, powers, and duties which, in its opinion, will tend

to prevent transportation accidents and promote transportation safety

(2) Conduct special studies on matters pertaining to safety in transportation and the prevention of accidents

(3) Ensure that in cases in which it is required to determine cause or probable cause, reports of investigation adequately state the circumstances of the accident involved

(4) Initiate on its own motion or conduct rail, highway, or pipeline accident investigations as the Board deems necessary or appropriate

(5) Make recommendations to the Secretary or Administrators concerning rules, regulations, and procedures for the conduct of accident investigations

(6) Request the Secretary or Administrators to initiate specific accident investigations or conduct further investigations as the Board determines to be necessary or appropriate

(7) Arrange for the personal participation of members or other personnel of the Board in accident investigations conducted by the Secretary or Administrators in such cases as it deems appropriate and

(8) Request from the Secretary or Administrators notification of transportation accidents and reports of such accidents as the Board deems necessary.

Sec. 5 (e) Except as otherwise provided by statute, the Board shall make public all reports, orders, decisions, rules, and regulations issued . . . and the Board shall also make public--

(1) Every recommendation made to the Secretary or an Administrator;

(2) Every special study conducted; and

(3) Every action of the Board requesting the Secretary or an Administrator to take action, ...[2]

In 1966, the President, in his message to Congress proposing the creation of the National Transportation Safety Board,[3] pointed out the inadequacy of our knowledge of why accidents were occurring, the fact that existing safety programs were widely dispersed, the absence of a clear assignment of responsibility for transportation safety at the Federal level, and the inadequate allocation of resources for safety. The President urged a strengthening of the Federal role to give greater support to safety programs, the development of a comprehensive program keyed to total transportation system needs through safety techniques proven in one mode to be applied to others, and better

TABLE 3

Recommendations Made by NTSB and Actions Taken

Recommendation	Action Taken
General Surface	
Centralized reporting system be established within DOT and that a data clearinghouse be established for 24 hours a day immediate dissemination of information on control of hazardous materials involved in accidents.	DOT special Hazardous Materials Task Force concurred in recommendations. DOT has issued proposed regulations for the reporting of hazardous materials accidents.
Establish system safety engineering approach within the Office of High-Speed Ground Transportation and develop in-house capabilities to administer and supervise system safety programs in high-speed projects.	Engineer assigned to carry out recommendation.
Highway	
DOT consider requiring seat belts or other restraining devices for all persons in buses.	Notice of proposed rule-making issued requiring seat belts for drivers of buses.
DOT develop methods whereby hazardous driving conditions, such as fog, can be brought to drivers' attention.	The Bureau of Public Roads and the Oregon State Highway Department conducted a research project in Oregon involving the use of patrols, special lights and signs and warnings to drivers. Results of that study and another one being conducted by the National Cooperative Highway Research Program of the National Academy of Sciences are under review.

Recommendation	Action Taken
Improve the placarding requirements for motor vehicles carrying hazardous materials.	Regulations changed by DOT to comply.
DOT initiate full-scale studies of the hazards and sources of potential catastrophe involved in the use of highway airspace.	Copies of NTSB study distributed widely by Bureau of Public Roads asking for suggestions and comments.
Rail	
Regulations to require immediate reporting by railroads of accidents involving hazardous materials.	A notice of proposed rule-making has been issued to establish requirement.
Additional railroad safety legislative authority be sought if accident problems not solved by railroad industry.	DOT submitted Railway Safety Bill presently being considered by the Congress.
A railroad make a reappraisal and self-assessment and take corrective action on inadequate rules, operating practices, personnel training, ineffective organization, inadequate use of modern communication devices, and lack of application of modern techniques in human engineering, which had caused a major railroad accident.	The Company has revised its operating procedures and its rules.
American Railway Engineering Association revise its track and maintenance standards.	The Association of American Railroads issued new track and maintenance standards.
Special examinations be conducted of all train wheels manufactured by a corporation to check for rough machining and poor finish and defects, which had caused a major accident.	Association of American Railroads (AAR) issued a bulleting stating that all such wheels are to be given a detailed inspection when railway cars are in the shop.

Recommendation	Action Taken
DOT establish a Task Force, including representatives of industry, to study the feasibility of a device for application to the front end of locomotives to cushion and deflect motor vehicles in collisions.	DOT agreed with the recommendation and referred it to the Joint Action Group on Grade-Crossing Safety where it is now under study; contracts also awarded to study; AAR studying.
Rail-Highway	
State and local schools governing bodies include schoolbus routing to avoid grade crossings for school bus operating safety.	Twenty-seven States have acted on recommendation and have incorporated recommendation in operating procedures for school buses.
DOT seek legislation to authorize prescribing regulations requiring emergency escape means for railroad passenger cars and emergency lighting for railroad passenger cars, and emergency brake release in cab of trucks to permit vehicles to be moved to safe location off grade-crossing tracks.	The proposed Railroad Safety Bill mentioned above includes authority to implement these recommendations. Latter recommendation under review by DOT.
Study questionable audibility of external sound signals within motor vehicles to ensure reliable reception.	Contract awarded to study this subject.
Pipeline	
DOT require current reporting by state agencies and industries of pipeline accidents.	Recommendation adopted.
Marine	
Coast Guard conduct detailed special inspections of all Great Lakes cargo vessels and upgrade	Coast Guard conducted series of special inspections of all Great Lakes cargo vessels, and

Recommendation	Action Taken
the safety conditions of such vessels on a ship-by-ship basis, to avoid repetition of accident.	extensive repairs and upgrading have been completed.
Regular and special weather warning information be widely disseminated by all means of communication to warn boaters of dangerous conditions, to avoid repetition of loss of and damage to motorboats with extensive loss of life in Lake Michigan.	State of Michigan, through local police departments and sheriffs' offices, has complied. Since the date of this accident and despite a marked increased in recreational boating and fishing, there has been no repetition of this type of accident in this area.
Coast Guard revise its regulations to ensure demonstration of proficiency by Merchant Marine Officers in the use of radar.	Coast Guard has revised its regulations for examination in connection with the issuance of original and renewed licenses.
To avoid repetition of major marine accidents due to improper handling of hazardous materials, Department of Labor establish a training program in stevedoring organizations with emphasis on hazardous materials and revise regulations prescribing the number of items which can be placed on a single pallet and methods of securing such items while being moved from the dock to the ship or vice versa.	Department of Labor changed its regulations to comply.

Aviation

Recommendation	Action Taken
After fatal accident revealed loss of fuel to the engine due to inadequate tank and plumbing design, design changes to provide a dependable fuel supply for that model aircraft were recommended to FAA.	Airworthiness Directive issued requiring placarding all in-service aircraft of that model, i.e., fuel flow problem, and advising of a retrofit fuel tank being available and requiring all new aircraft coming off the production line incorporate the newly designed tank.

Recommendation	Action Taken
FAA promote development and installation of audible and visible altitude warning devices, to prevent accidents due to misreading and missetting.	Rule adopted requiring minimum decisional altitude alerting devices in jet aircraft by February 20, 1971.
Reassessment of altimetry systems with regard to their susceptability to interference by forms of precipitation.	FAA, the aircraft industry, and NASA have allocated resources and are presently making such assessments.
Development of microwave pictorial display approach aids as a possible improved replacement for the barometric altimetry system.	FAA evaluating military microwave pictorial display program for application as additional approach aid in civil avaition.
FAA reevaluate criteria and procedures affecting a change in design and/or rework of aircraft components critical to continued safe operation be undertaken to upgrade standards and technique where necessary, to present inflight failures of reworked main rotor blade spindle of a commercial passenger helicopter resulting in fatal accidents.	FAA initiated review of procedures used by manufacturers, operators, and repair stations, in categorizing and evaluating major and minor repairs and design changes, and reviewed the procedures, methods, and personnel qualifications of the repair station involved in the particular accident and instituted at repair station procedures to upgrade the inspection methods and rework and design change techniques for a safer product.
FAA initiate an educational program for pilots and air traffic controllers, i.e., midair collision problem.	FAA is preparing a film on the midair collision problem and plans to emphasize the importance of pilot vigilance and collision avoidance in all proposed training material and advisory circulars.
FAA expedite development of low-cost collision avoidance systems for all civil aircraft.	FAA now zeroing in on the collision avoidance hardware which shows promise for immediate application.

Recommendation	Action Taken
FAA conduct detailed traffic flow study for all high-volume general aviation controlled airports to improve Visual Flight Rules traffic flow techniques of Air Traffic Control to prevent midair collisions.	Study to chart VFR routes and check points underway.
Designate climb and descent corridors for high-performance aircraft at high-density airports.	FAA exploring corridor concept at lower-activity airports.
FAA review in depth manufacturing process and quality control system of a manufacturer whose product caused a major aviation accident, to ensure effective quality control.	FAA convened a Production Certification Board to perform an in-depth investigation; manufacturer was requested to make significant changes in both the manufacturing methods and quality control procedures.
FAA alert users that laboratory examination of two failed bolts causing an accident showed threads had not been formed by rolling and had a lower fatigue strength than required by specification.	FAA issued Airworthiness Directive requiring a periodic inspection for proper bolts and a periodic dye penetrant inspection of these type bolts for thread cracks.

coordination between cities, states, and the Federal Government looking toward a partnership to solve existing problems.

President Johnson described the Safety Board as an agency whose "sole function . . . will be the safety of our travelers," and added that the Board would "review investigations of accidents to seek their causes, . . . determine compliance with safety standards, and examine the adequacy of the safety standards themselves."

The Chairman of the Senate Commerce Committee, Senator Warren G. Magnuson, described the Safety Board as "the Supreme Court of transportation safety," while a Senate committee stated that "the Board was to play a significant and meaningful role in and serve as the focal point in the Federal establishment, for transportation safety. . . ."[4] A House committee said it would be a "key agency for energizing the departmental offices and agencies engaged in accident investigation and prevention work."[5]

The National Transportation Safety Board is headed by five members appointed by the President and confirmed by the Senate. Unlike the other functioning segments of the Federal Government that were brought together under the Department of Transportation, the new Safety Board is autonomous. The Act states specifically that the Safety Board, in the exercise of its functions, powers, and duties, shall be" independent of the Secretary [of Transportation] and other offices and officers of the Department." Furthermore, the Safety Board is directed to report to the Congress annually on the conduct of its functions under the Act and also to make recommendations for legislation deemed appropriate.

The Safety Board is charged with a continuing review of general safety in all modes of U.S. transportation. It exercises broad safety powers over marine, railroad, highway, and pipeline functions, and over all civil aviation. In surface accidents, it makes the final determination as to cause in any case it elects to investigate or review. In civil aviation, it must determine the cause of all accidents.

In safety matters, the Board has relationships with the Federal Aviation Agency (FAA), the Federal Railroad Administration, the Federal Highway Administration, the Coast Guard, the National Highway Safety Bureau, and the Office of Pipeline Safety, which are the primary agencies of the Department for regulating safety in transportation. For example, the FAA operates the nation's air traffic control system, certificates airmen and aircraft, and promulgates the Federal Aviation Regulations which govern U.S. civil aviation. But the Board has a continuing interest in the results of accident investigation for the purpose of determining what recommendations it can make to correct any matters which might have been shown as probable causes in accidents. To complete the circle, the FAA makes use of accident data developed by the Board in furtherance of the performance of its functions.

In the aviation accident prevention field the Board sits as an oversight or "stimulating" group, monitoring the activities of the FAA, the industry, and all those involved in air safety. Following an accident, if the Board determines that there are weaknesses in the system, a need for modification of aircraft or regulations, it sends specific safety recommendations to the FAA or others, spelling out what it believes should be done.

In surface cases, the Board operates similarly, leaving accident investigation to the Administrations and the Coast Guard. By delegation it has relieved itself of the responsibility of issuing reports and determining cause in most accidents. It reserves to itself, however, the authority to determine cause and issue reports in major accidents, hopefully not more than a total of 30 a year out of all the rail, highway, marine, and pipeline accidents of interest to the administrations of the Department. By such selectivity, the Board is better able to devote detailed analysis to these few most significant cases and learn lessons culminating directly in accident prevention and safety promotion, rather than attempting to undertake an extensive program, which would dilute its efforts due to the sheer job of on-the-ground investigation.

Although DOT provides many of our administrative and housekeeping services, the Board maintains an arm's length relationship with DOT. The Board considers itself as an arm of the Congress, and of the President. The Board maintains this arm's length relationship in order to make sure that there are no appearances of pressures that could be applied by the Department or that give the appearance of a conflict of interest by being too close to the Department. The Board appoints its own personnel, issues its own Annual Report to the Congress, and has direct access to the Congress. The Board is part of the Executive Branch of the Government, it is part of the Administration, and it must clear its budget and comments and proposals on legislation with the Bureau of the Budget. But in every aspect, the Bureau of the Budget, the Executive Department, and others promote the same thing the Board seeks: transportation safety.

The Board looks at specific accidents for defects in the technology of aviation, railroad, highway, truck, bus, pipeline, and marine transportation. It assesses the operational, technical, and engineering adequacy of such transportation. When investigations are completed, findings, conclusions, probable cause, and recommendations are extensively reported by the Board.

The Safety Board is concerned with individual accidents for another reason. Although statistics are significant because they tell us how important a certain factor in accidents may be, we must know first what factors are involved in accidents--in other words, what kinds of things ought to be measured by other investigators and what data reported to and collected by the statistician. When we find a technical factor in accidents, we must determine what contribution it made to the accident. Have we learned something which may often be repeated in other accidents? Then we should collect the same or similar data as a basis for determining their implications. But, these truths simply cannot be found by statistical analyses alone.

The Board does not necessarily wait for individual major accidents to occur before it acts. It often initiates special studies on the basis of a series of accident reports with similar characteristics and some indication of trends and on the basis of statistics drawn from the history of accidents with similar or related characteristics, which possibly indicate trends. Such studies include findings, conclusions, and recommendations for preventing similar accidents. The titles of special study reports issued during calendar year 1969, are as follows:

1. Application of System Safety Concepts in High-Speed Ground Transportation System

2. Application of New Impact Technology to Head Ends of Locomotives

3. A Study of Safety Considerations in the Utilization of Airspace Over and Under Federally Aided Highways

4. Alcohol Problems and Transportation Safety

5. Uniform Reporting System for Accidents Involving Shipment of Hazardous Materials

6. Compatibility of Standards for Drivers, Vehicles, and Highways

7. Recreational Boat Accidents, Boating Safety Programs, and Preventive Recommendations

8. Towing Vessel Safety and Accident Prevention Recommendations

9. NTSB Study of Midair Collisions in U.S. Civil Aviation--1968.

The Board does not believe it necessary to completely solve a safety problem before it makes a recommendation. Sometimes it considers it adequate to identify the problem and say, "Look, something has to be done about this." Some of the studies listed above are in this category. Many of them relate to technology assessment, and the recommendations flow from assessment of the currency, value, and contribution to safety of the technology of the transportation industry.

Sometimes the Board finds it best to report to the public on safety problems and possible solutions through the medium of a public hearing. For example, the Board selects an accident or a series of accidents or a special and serious safety problem that occurred on a railroad, a highway, or a pipeline, and holds a hearing introducing into the record all the evidence found, and invites parties-in-interest and witnesses to provide additional information, if available. Recently the Board made a study (#9 above) of midair collisions and found that, in addition to the special study report it issued, there was a serious enough problem to hold a hearing. It then invited representatives of the industry, manufacturers, various associations, the Department of Transportation, and others to testify about what could be done about preventing midair collisions. A report will eventually be issued, summarizing the findings of the Board in the hearing and making recommendations. However, because of the immediate need to correct certain problem areas the Board will not wait until its final report is published, but has already proceeded to issue four recommendations with respect to midair collision safety aspects developed at the hearing.

The Board has also recently held extensive hearings on a chain-pile-up accident with many fatalities on the New Jersey Turnpike, and following, on a similarly tragic and serious natural gas pipeline accident in Gary, Indiana.[6] It is now holding a hearing on a scheduled air taxi accident with multiple fatalities.

It should be reiterated that the Board moves quickly to make recommendations to prevent accidents and does not wait until the report on a matter is completed. With lives, as well as other important issues at stake, the Board makes its recommendations as soon as it recognizes any possibility for improving safety or preventing a similar accident.

In summary, the following is stated in our 1968 Annual Report to the Congress:

> But, before all safety solutions can be found, there must be identification and definition of the problems. The Safety Board has sought to assist in this effort by using lessons learned from in-depth investigations of selected surface accidents, by emphasizing safety promotion efforts in surface transportation, and by strengthening its accident prevention efforts in aviation as a logical follow-up to its continuing accident investigation efforts, by urging more comprehensive accident investigations by the Administrations of the Department of Transportation and by states and local communities; by urging cross-modal application of accident investigation techniques which have proven to be successful in one mode; by highlighting, through statistics, the seriousness of the railroad and recreational boating accident problems; by making special studies and using them as a springboard for safety promotion; and by making recommendations to the Department of Transportation, to the transportation industries in general, to the public, and to representative organizations and associations concerning safety problems and how to cope with them in all modes of transportation.[7]

Safety problems will continue to escalate under the pressures of population growth, the introduction of new and many more vehicles into our congested facilities, and changes in technology and related (and phenomenal) increases in speed in all modes of transportation. In aviation, the introduction of the supersonic transport and jumbo jets and the continued dynamic growth of general aviation will pose new and complex safety problems. The other modes will face similar problems. The number of buses, trucks, and automobiles especially will continue to rise. Problems in railroad safety and grade-crossing safety will be accelerated with the introduction of high-speed trains. Transportation systems now on the drawing boards will present new, complex, and challenging problems. With these new developments on the horizon, the Safety Board believes it imperative that all steps now being initiated to assist in accident prevention be pursued with vigor and expedition. More must be learned about what causes accidents so that significant accident prevention steps can be initiated.[7] Under this concept, potential accident causes must be identified, starting with the nucleus of an idea for an advanced transportation system. New safety techniques must be ready for use as new methods of transportation appear on the scene.[7]

Board recommendations to the modal administrations go beyond the specifics of one accident. They may involve how investigations may be better conducted; which rules or regulations should be amended (e.g., construction of vehicles, use of seat belts for bus passengers, better construction of school bus shells, training of railroad personnel) which may affect the technology of an industry; how Federal funds may be better used; what further steps may be taken to improve industry procedures and operations. Sitting as an oversight unit, the Board, one step removed from the day-to-day and regulatory activities, is able to look at accidents in the mass as well as individually and point out areas of needed accident prevention attention, which might understandably be overlooked by persons who have the close-up day-to-day working responsibility for safety.

The Board would like to believe that the industries, the carriers, and others will see their own self-interest in improvements in technology through the cutting down of accidents, injuries, and economic loss and always hopes they will do something about accident causes at least after their attention is called to them by Board recommendations. But the Board does not hesitate to come back a second time and recommend legislation or some other more specific action requiring the carrying out of the recommendations. Although the Board cannot force the carrying out of the recommendations, it can suggest to others that they put the pressure on--whether this be the Congress, another Government agency, the public, or the press. For example, if we make a recommendation that a part of an aircraft or an operating practice be changed, and the FAA carries out the recommendation by advising (but not requiring) a manufacturer or carrier to take action, but this doesn't do the job, the Board does not hesitate to come back and tell the FAA the requirement should be made mandatory.

On occasion, safety recommendations--especially those criticizing current technology or technology not adequately reactive to need for change or improvement, thought by some to be desirable--run into the hard-core facts of economic life. Some run into resistances of other types, some warranted, some not. Pride has its effect as does "nih," i.e., "not invented here." Such situations may involve some type of compromise. There may be differences of opinion on occasion, as to a particular decision where conflicts exist; but the important thing is whether a reasonable accommodation can be reached.

Joseph J. O'Connell, Jr., former Chairman of the Board phrased it thus:

I might say, in passing, that the main reason for the

> existence of the Safety Board could be summed up in a saying, which is to the effect, and I quote, "We all look at the same things, but rarely do we all see the same things." I need not develop this point further, other than to observe that a major thought behind the creation of our Board is that, one long step removed both from the operational problems of the industry and the direct obligation to regulate it, a board such as ours should be operating from a point of vantage which will enable us to see, with an objective but not jaundiced eye, matters and things to be done not adequately perceived by people of equal goodwill but otherwise preoccupied.[8]

When the Board does engage in investigations of surface accidents--and it does, on occasion, hold hearings--it comes in close contact with the industry. The Board requires its staff to keep aware of what new techniques are available for safety and accident prevention; it follows these developments within the industry; it examines current and prospective practices, rules, and procedures; and from all the above, assembles reasonable conclusions and effective safety recommendations which it feeds back to further stimulate and advance safety consciousness developments and improvements on the part of the industry and others.

In one sentence, the Board's greatest contribution to the cause of transportation safety is in pressing for development of better accident prevention programs, rather than in being a post mortem investigative agency. For example, at one point early in its career, the Board called attention to what should have been obvious: the fact that hazardous materials are being transported in ever-increasing large volumes, giving rise to greater possibilities for catastrophic occurrences should nonsafe transportation practices permit accidents to occur.

More specifically, in early 1968, the Board noted a serious increase in railroad accidents: derailments had increased 100 percent over a period of about seven years. Although deaths, economic loss, and injuries increased due to the accidents, which is bad enough, a simple derailment of cars carrying combustible, explosive, corrosive, poisonous, gaseous, or atomic substances could make a minor derailment a great catastrophe, entailing possibly thousands of deaths and injuries and many millions of dollars of damage. The Board has investigated, held hearings, and issued reports in a number of such derailment catastrophes where as many as ten city blocks were leveled, water supplies poisoned, and persons killed or injured. If such accidents had happened in more populated areas, hundreds or thousands might have been killed.

The Congress has asked the Board if this power of recommendation, without supplementary authority, is strong enough to obtain results. So far, the Board believes that if its recommendations are reasonable and sound, then the recipient will be cooperative and willing to comply; and if he doesn't, he will ignore them at his peril. The Safety Board wishes its recommendations to attain acceptance on their judicial and technical merit. On the other hand, the Congress has already given the Board a certain amount of "supplemental authority" since it required that the Board make all recommendations public. Well-thought-out recommendations to improve safety in transportation that are known by the public will be supported by the public. The self-interest of the party concerned is used to make changes. For example, industries are accepting our recommendations (in terms of joining us rather than fighting us), if our recommendations make sense.

New authority to require mandatory acceptance of its recommendations might put the Board into a regulatory position, and this might place the Board in a compromising position. For example, if an accident occurred which involved a regulation or requirement that the Board required to be imposed, could the Board satisfy everyone that in its investigation of that accident it could or would objectively determine whether that Board regulation or requirement was a good one, may have contributed to the occurrence of the accident, or may require change? It appears most desirable that the Board maintain a "hands off, arm's-length position" with respect to regulation, law, enforcement, etc., so it can be as objective as humanly possible.

The Board makes current assessments and decisions based on such assessments, which are reflected in findings, conclusions, and recommendations, and a feedback process results. This provides not only a basis on which to determine the effectiveness and value of the recommendations, but it also provides for continual assessment of how the assessment and recommendation process might be improved.

For example, the Board regularly follows up to see what is done about the recommendations; this leads to a review of techniques for making the recommendations, their support, and their strength. The Board's procedure has ranged from short recommendations to those supported by a lot of detail; has made recommendations affecting an individual carrier, an individual manufacturer or plane, an entire industry, a whole group of planes, or a range of types of vehicles; has compared systems (for example, compatibility of the driver, the highway, and the vehicles); has made recommendations on parts of airplanes, parts of vehicles, or parts of trains; and made recommendations on the need to apply system safety as a transportation development concept, for example, for high-speed ground transportation developmental activity.

Recommendations are repeated in speeches, highlighted in reports used in testifying before the Congress. The Board learns, from its own assessment process, how to do a better job of technology assesment and how to make better decisions and more effective recommendations. Being just three years old, the Board is still in an evolutionary process: even the format of the basic accident or study report itself is in a state of flux. Decisions are sometimes made on an ad hoc basis as to whether the most effective presentation, format, style, sequence, etc., is being presented.

The Safety Board can address its recommendations to Congress, any Federal agency, the industry, and others. Such recommendations are usually in the form of a letter signed by the Safety Board Chairman, which is placed in the Safety Recommendation Docket in our headquarters office five working days after mailing. The Safety Recommendation Docket is public and is reviewed in the Public Affairs Office by daily visitors. In addition, if the Board deems more or more immediate attention is warranted, it may issue a press release or otherwise call the recommendation to the attention of the news media. In some cases, the Board has written to state bodies, manufacturers, and carriers to call special attention to some of its reports and recommendations.

The form of the Board's final and only action, namely, the issuing of recommendations, as the best way to achieve improvements of a technical or operational nature, sometimes elicits unfriendly responses. However, Chairman O'Connell offered these alternatives:

> Of course, you will constantly be confronted with the spectre of Government regulation, both safety and economic; and in the safety field you may very well have to continue to put up with an organization such as I represent which, on occasion, will be impelled to make recommendations directed either to a segment of your industry, to a Federal agency having safety regulatory responsibility, or quite possibly to the Congress. Any one of these may not be well received, but I can assure you that it will be well intended.
>
> We fully realize that the major responsibility for the safety of your operations is that of your organizations. In that vein, I perceive our function as in major part one of encouraging industry programs of self-help, spurring them on, if you will, by such recommendations and suggestions as the Safety Board may feel impelled from time to time to make. It is also true, however, that a failure

> of any segment of the industry to achieve what is regarded by the public as an acceptable level of freedom from hazard invites more stringent regulatory mechanisms, including more legislation.
>
> The legislative alternative, and I do not consider it a basically desirable one, is for use or not depending upon how effective you police your own establishment.
>
> I am not one of those who believes that every time we discern what appears to us to be a weakness in the safety record or the safety program of a segment of your industry, the answer is necessarily more regulation or another law.
>
> In my view, it is entirely too simple to say that an observed weakness in an institution can only be, or best be, corrected by another law or regulation. Sometimes, in fact, a law or regulation decreeing a particular device or technique in the field of safety in transportation may even be illusory. It may even con an unsuspecting public into the belief that things are better solely by virtue of a statutory or administrative measure. You gentlemen know as well as I do that such is not necessarily the case.[8]

We come now directly to technology assessment, and honest-to-goodness examples of the Board's assessment process, and its influence on the development and application of means to achieve a better, safer, and more modern and effective transportation system. Although the Board's activities are safety-oriented, concomitant parallel results from its activities are improved efficiency, economy, comfort, and service. Examples of safety recommendations and the results of those recommendations issued by the Board since its origin in 1967 (which follow), highly selected out of many more, give insight into the Board's technology assessment activities. These recommendations evolve from investigations of specific accidents, from studies of a group of similar transportation accidents, or from generalized studies of conditions, situations, or statistics. These recommendations and their effects are cited in Table 3.

The recommendations of Table 3 are effective in upgrading parts of current systems or providing for improved technology in completely new departures, e.g., retrofitting versus completely new mechanics. Even though the recommendations are worded in safety terminology and are safety-oriented, they do consist of elements which have a positive effect on technological developments in aviation, rail, highway, pipeline, and marine transportation.

The new jumbo jets, the faster trains, the larger trucks and buses, the bigger pipelines, the containerized ships, the smaller automobiles, are now all being engineered with safety as a more important consideration than heretofore. Although the Safety Board recommendations are not mandatory, in the aviation area about 75 percent or more of the recommendations are accepted by the FAA and others; and in the surface area, so far about 50 percent have been accepted. Many of the remaining safety recommendations are "open" in terms of the recency of the recommendation, and undoubtedly a larger number of them will be accepted and carried out. Many of the Board's recommendations, as you can note above, evolve into changed technology. But, which is the primary or the by-product: increased safety or improved technology? improved operations? less economic loss, suffering, injury, death? improved facilities? greater comfort?

The Board is moving back into the accident process because it doesn't want to wait for the accident to happen; it wants to get back to the drawing board and even back to that point when the item is a first glimmer, so that system safety is instilled all through the process to prevent accidents.

The Board does not believe in simply reacting, but rather in anticipating--our objective is to prevent accidents. And this leads us into the prospective assessment of technology in the transportation industry. For example, the Board has been carefully following the development of the Boeing 747. The experienced loss of prototypes on a learning curve basis (e.g., several accidents in the first eighteen months of a plane's use) is patently unacceptable, and the Board has been and is carefully watching incidents (i.e., an occurrence short of an accident and which waves a warning flag) as a springboard for accident prevention recommendations.

Another example of assessment of technology is reflected in the Board's study on the compatibility of the driver, the highway, and the vehicle--leading to recommendations that more effective coordination take place among the different Government activities writing standards for these three areas. We also have examples of technology assessment to cover the air traffic control system, surface traffic control, and police accident investigations.

In addition to reviewing the activities of those who can do something directly about factors causing accidents, such as manufacturers or carriers, the Board looks at the work done by others who are charged or being paid to interest themselves in taking action on transportation safety, to see whether they do the job they are hired for. This may be considered to be a "second-guessing operation." Put another way, we are also the watchers of the watchers responsible for the transportation safety in all modes.

Many activities in the Department of Transportation, for example, have a direct operating, regulatory, and enforcement responsibility. It is the job of the Board to evaluate the activities of such units and issue recommendations for improvement. Are they doing the best job for the resources invested? What is the payoff in safety? is the investment of available resources, effort, and interest in balance among other efforts and also among the different modes of transportation? Are regulations issued contributing to or derogating safety? Are regulations up-to-date, and are they being implemented effectively? Is the regulation writing unit learning from experience; e.g., if a regulation for brakes is issued and accidents are caused (or not prevented), is the Government unit moving promptly to change the regulations? Does it learn only from accident experience, or does it anticipate? Are determinations of causes of accidents used only in retrospect (post mortem) rather than for projection by manufacturers, carriers, Federal units, state and local governments, associations and others, to propose or issue better regulations, to build better vehicles, and systems to accommodate them better; to train pilots and drivers better?

TECHNOLOGY ASSESSMENT AND VALUE OF NTSB WORK

The operations of the National Transportation Safety Board are a living example of technology assessment. It hopefully makes friends, primarily among the traveling and shipping public, but its objective is to change the technology of transportation to improve safety and prevent accidents. The Board relies entirely on its authority to issue recommendations and get them accepted by responsible people, including Congressional committees, the industries, governmental agencies, manufacturers, carriers, on the basis of their being worthy of implementation.

It is difficult to measure the Board's effectiveness in terms of cost benefits. There are some examples in the aviation area. For example, Board accident investigators found a portion of a wing had come off a large plane, causing many deaths, due possibly to a fatigue fracture. The Board recommended to the FAA that all planes of this type be inspected forthwith; they were, and eight others with evidence of similar fatigue areas were found. Corrections were made forthwith.

We would like to think that this did avoid additional accidents.

The flight recorder was originally designed by personnel in the NTSB to assist in aircraft accident investigation. The flight recorder is a device that records altitude, heading, speed, and vertical acceleration against relative tape time. The Board, other Government agencies, and the aviation community have developed it further for many practical uses. The maintenance recorder, a new apparatus grafted onto the flight recorder, adds a greater number of channels to gather data to provide for proper maintenance, and thus safety. It provides real time information on when to replace or repair engines or parts. It will reduce equipment down-time, reduce cost of repair work, and promote more efficient management of maintenance activities. This device is, by its very nature, an accident prevention device It is both cheaper and safer to remove an engine because of tell-tale warnings than to have it remove itself in flight and, in the process, break primary structure and kindle a fire. This type of recorder can also be used in pilot training or in evaluating pilot performance. A review of the flight in all the various parameters can show pilot weaknesses.

An established aircraft manufacturer rarely produces an entirely new aircraft. For practical and economical reasons, many components and systems of their present model aircraft are incorporated in the new aircraft. Therefore, the Board's findings of poor or inadequate design of materials in the present aircraft help preclude them from being used in new generation aircraft.

The Board is, of course, public-interest-oriented, but you might say that that is also true of all three branches of the Government. At any rate, whether others are or not, we view ourselves as representing all aspects and all entities in striving for greater transportation safety and producing improvements in technology that result from most of the striving. Transportation safety is really a consumer issue; deaths, injuries, and economic loss to carriers and other manufacturers are no small matters. They directly affect everybody's pocketbook in addition to affecting those whose families or relatives are killed or injured. Everyone who travels or uses any transported product (and is there anyone who doesn't?) must contribute to the cost of unsafety.

The Board activities contribute to improving the quality of life. Transportation safety has deteriorated with increases in the number and severity of accidents due to increases in the number and speed of automobiles, railroad accidents, pipelines being built nearer to homes or homes being built closer to pipelines, large hazardous materials movements by water, and other technological developments. The rates must be reversed significantly. It isn't only the accident causer who suffers--it's too often the innocent bystander. The Board also views itself as a user of "seed money," i.e., requiring only a small investment, it energizes others to do their jobs better, with their larger resources. For example, the Board is interested in reviewing the balance of effort and resources applied by Government agencies and others, to rail, highway, pipeline, marine, and aviation safety matched against the incidence of deaths, injuries, economic loss and potential catastrophes. Further, it is concerned whether enough effort is being applied in a balance method to reduce accidents to the maximum extent possible?

For example, if a Government agency has a budget of several million dollars, should it spend its funds primarily to enforce safety regulations by investigating compliance and seeking prosecution of those who do not comply; or should a larger share of available funds be spent for research or for doing a better job of investigating accidents to determine these causes so corrective action can be taken more promptly? If aviation accidents cause 1,500 deaths a year and a half-billion dollars is invested in investigating aviation accidents, what should the investment be for highway safety, when almost 60,000 lose their lives each year? In its recommendations, the Board points the direction, based on real life experience of how best to invest available funds for safety and where greater resources or resources in better balance are needed.

The public is becoming more aware of our activities, and we are getting more and more letters asking us to do more things. We are becoming known as the consumer representative agency for transportation safety and are being invited in by state and local governments to conduct accident investigations or studies. We have had to fend these off because we would soon become involved in so many local interests that we would soon be in some difficulty in trying to solve many individual problems rather than broader and generalized problems.

I believe our approach, finding large areas of safety concern and urging correction of them by recommendations, is the best way to get the job done. The Board is probably never going to run out of business. There are so many matters affecting transportation safety and related technology that we have to set priorities on what we're going to do, and we obviously are in no better position than other Government agencies in getting funds for what should be done. On the other hand, we don't want to get too big or develop too much of a self-interest in growing. Consequently, we set priorities and have plans for reviewing areas of technology related to safety in transportation which go into the next several fiscal years.

Inherent in the Board's activities is an aspect of duplication. In some respects and probably in many aspects, the Board duplicates the work of others. As the Board looks at a selected number of accidents, or makes special studies, it is duplicating or repeating what others are doing, or are charged with doing. But it is "due duplication." Other entities in government, industry and elsewhere strive to make the technology of planes, locomotives, trains, cars, buses, trucks, contribute to safety, but they are under pressure of other contravening interests. A Government agency issues regulations pursuant to administrative procedure rules and somewhat based on subjective judgments; and the carriers and manufacturers have to pay for the cost of safety in the light of competitive pressures.

The Board is cast in the role of reviewing all these considerations objectively, perhaps covering the same or similar ground in our studies and investigations. But this is necessary. A parallel is the review of certain activities of the Executive Department by the Congress; again "due duplication." We see ourselves in the same light; but, as contrasted to Congressional committees, transportation safety is the Board's only business.

The Board is discovering or rediscovering many obvious things. For example, it should have been obvious to many that the technology of hazardous materials has a direct relationship to accidents and safety (there is a new exotic chemical developed or invented every hour on the hour, actually twenty-five a day, and for many of these we don't know all their characteristics when they are initially shipped on special permits). This presents, literally, an explosive situation in transportation. Yet the Board found it necessary to again and again call attention to the danger of moving hazardous materials because of the possibilities of increasing numbers of accidents leading to major catastrophes. On the other hand, hazardous materials are useful to the quality of life and economic advancements. No one wants to stop their movement entirely. But the Board called for a greater awareness of the dangers and how to handle them when an accident occurs.

Whether the Safety Board concept should be exported and similar agencies formed to cover other areas on a continuing, ongoing organizational basis (rather than an organization which prepares a report and then goes out of business) is an interesting question. For example, for programs concerned with environment protection and improvement, social problems, crime, consumer product matters, other types of safety, urban problems, it might be worth considering having an organization such as the NTSB oversee how others are doing their job, because there may be conflicts, duplications, and perhaps even oversights and issues and matters falling between the cracks.

Perhaps some of the new committees and commissions that are being established on a temporary basis by the President or by legislation to cover specific problem areas are similar, but there are also differences between what the Board does and how it does it and the above-described short-lived commissions and committees. The Board is established as a continuing, ongoing activity; some of the special committees and commissions issue a report and then go out of business. There are undoubtedly advantages to both approaches. Some problems may just need a report for action thereon by the Executive or the Legislature. Continuing problems such as transportation safety require a continuing review body which keeps in there pitching, or as Safety Board Chairman John H. Reed said in a recent speech: "Work in transportation safety is never done--of necessity, it is a continuing process that must never end."[9]

In summary, the Board's interest are the interests of all who benefit (or suffer) from adequate or safe transportation--in terms of where we have been, where we are, where we are going. I believe the Board is a desirable form of Government activity, considering the fact that we're losing about 60,000 people a year on the highways, more than 2,000,000 injured, in more than 14,000,000 accidents, with an economic loss to the community in excess of $11 or $12 billion a year. Add to that the losses in aviation, marine, and pipeline activities; and add the potential for increases in risk as the number of cars and the amount of movement of materials and people grows; and add the potential for catastrophe. I believe that the best regulation is the least regulation, with minimum controls to improve the quality of life, yet balancing of interests: public, economic, governmental. With proper attention, the safe movement of people and their cargo can be attained within reasonable costs and safeguards. The Safety Board's budget is about $5 million annually. One Boeing 747 costs more than $20 million; one transportation-related catastrophe could exceed a billion dollars.

NOTES

1. U.S. Senate, Committee on Government Operations, Hearings on S. 3010, 89th Cong., 2nd sess. (March 29 and 30, 1966).

2. P.L. 89-670, 80 Stat. 931 (1966).

3. See President's Message to the Congress, March 2, 1966.

4. U.S. Senate, Committee on Government Operations, Report No. 1659, 89th Cong., 2nd sess.

5. U.S. House, Committee on Government Operations, Report No. 1701, 89th Cong., 2nd sess.

6. National Transportation Safety Board, Pipeline Accident Report, Low-Pressure Natural Gas Distribution System, Gary, Indiana, June 3, 1969 (December 4, 1969).

7. National Transportation Safety Board, Annual Report to Congress (1968).

8. Joseph O'Connell, Jr., "Safety--An Indispensable Ingredient of Any Sound Transportation System," an address before the Board of Directors of the Transportation Association of America, New York City (January 30, 1968).

9. John H. Reed, "Safety Trends for the Seventies," address before the American Trucking Associations, Inc., 1969 Newspaper Safety Writing Competition Awards Ceremony, Washington, D.C. (April 9, 1970).

DISCUSSION OF PROCESSES OF TECHNOLOGY ASSESSMENT: THE NATIONAL TRANSPORTATION SAFETY BOARD

Mr. Weiss introduced the discussion by stressing the uniqueness of the NTSB. In short, the NTSB is an "organization primarily involved in accident investigation and study . . . that makes recommendations with respect to these accidents." The studies are designed to "find what we think are faults in the transportation system which lead to lack of safety." Mr. Weiss continued:

There are some other organizations which have similar

> techniques to ours but one of the major differences is the fact that we have . . . absolutely no authority. The only thing we can do when we find the cause of an accident . . . is to make recommendations; those recommendations must, by law, be public.

The recommendations of the NTSB are ignored "at the peril" of the group that receives them because "there's always a darn good chance that we are right." The recommendations of the Board have found wide acceptance; 75 per cent of recommendations made to the FAA are accepted, for example.

The NTSB desires no power or authority beyond the issuance of recommendations, Mr. Weiss said. Any regulatory authority which the NTSB might receive would only destroy the "independence we have because we would . . . then become self-interested in our own regulation and we wouldn't be objective." The preservation of independence, he noted, is crucial to the success of the Board. Further, not only must the Board _be_ independent but it must maintain a reputation for independence _and_ objectivity: "The appearance of nonobjectivity or conflict of interest is just as bad as being nonobjective . . . . The greatest problem is to avoid being accused of being in favor of anybody."

NTSB AND ITS OPERATION

The apparent success of the National Transportation Safety Board leads quite logically to the question of whether some or all of its characteristics can be applied in other areas of technology assessment. Some participants seemed to feel that the operations of the Board cannot be generalized. One, a Government official, felt that in undertaking technology assessments one must be concerned with the "possible accessibility of a problem to modification." The NTSB can do an adequate job because "the railroad industry, the airplane industry, and the shipping industry are all very tightly organized and regulated things." He expected that the successes of an investigation group such as NTSB would be "greatest in those areas of society which are most organized."

Another participant expanded on this point and on the general question of how much authority an assessment entity should have. He asked:

> . . . isn't it a fact that the NTSB can be in a position of

> having no authority because there are so many other agencies of government that have regulatory authority whose buttons it can push? The transportation field is ideal; NTSB can be independent because there is an FAA and so on. But in many of the other areas of technology assessment we have no government institution filling the role NTSB is relying on someone else to play. I wonder whether you can . . . generalized the value of the Board's function--that is, divorced from regulation--where you have no regulatory authority anywhere in government.

One of those present noted that the Board was established to provide an independent entity which would be divorced from the promotional aspects of the transportation system. But since recommendations for safety improvements are made to manufacturers and government offices with promotional interests, this speaker raised some questions about the effectiveness of the recommendation procedure. He suggested that safety engineering might be done by "neutral parties." If this is not done, then "you introduce the conflict that the NTSB is trying to avoid" at a different stage in the examination of transportation safety.

A participant noted that the Board has been making practical efforts toward improving safety, but raised the issue of the difference between "doing what is practical and doing what is necessary, which may not be practical from a political point of view." He felt that the NTSB had been successful because it had stayed away from political "hornets' nests" and questioned whether it could continue to avoid touchy political problems in the future. He used the example of automobile safety to illustrate his point:

> We have pretty well milked dry the things that you can do that have high payoff by way of hardware improvements to the car. The easy things and the quick things that everybody thought about have largely been done. In the long term we must tackle the major job, which is . . . putting some constraints on the driver.

This task will not be as easy as putting new hardware on automobiles and may, in fact, require changes in the mode of operation of the Board:

> I'm simply saying that if technology assessment in the anticipatory sense is looking toward a longer-term future, then one of the questions that we have to ask

> ourselves is this: Is a mechanism which is quite satisfactory for this time period adequate to the problems that we are going to face in the future? And I'm suggesting that those problems . . . are going to require the constraining of the public, which means that you are in a political arena. I think there is reason to question whether the insulation, objectivity, and "recommendations only" philosophy of the Board will be functional in the future.

TECHNOLOGY ASSESSMENT: GENERAL COMMENTS

A university researcher noted that the safety area was one in which technology assessment could make positive contributions to the development of new technology. Technology assessment has often been criticized for being an essentially negative enterprise. But emphasis could be placed on

> . . . the possibility of new innovation which would bring forth safety improvements that could be extremely significant. There should be a way that the experience of the Board can be used to draw together some guidance that would help more effectively to orient research and development toward giving us technological innovation which would contribute to safety.

Another participant was concerned with both the interpretation of standards and the proper role of government in upgrading the safety of transportation systems (in this case, automobiles). He raised the possibility of

> . . . the Government using its power as a consumer, not as a regulator, to get some results in excess of the minimum. The Government buys a large number of vehicles each year, and one real possibility is for it to say that it will order cars from the manufacturer that provides the best safety package, over and above the minimum standards. The problem I'm addressing is: how do you avoid the problem of minimum standards becoming maximum? And the Government role as a consumer is one possible vehicle. At the very minimum we all expect as citizens that our Government will use our tax money by being as shrewd and tough a bargaining consumer as possible. The problem now is that the General Services

Administration says that since all cars meet the minimum standards we will buy any car, and the minimum becomes the maximum.

CHAPTER

11

TECHNOLOGY ASSESSMENT AND CITIZEN ACTION

Ellis R. Mottur

INTRODUCTION

It is appropriate to commence the last paper in this volume with a recapitulation of the points of consensus which emerged in the course of the preceding discussions.

First is the widely accepted assumption that it is extremely important for society to find ways of undertaking technology assessments far more effectively and expeditiously than has been the case in the past. We know that technology assessments have been carried out, albeit unconsciously, from time immemorial; further, even more official, professional assessments have been with us throughout much of the industrial revolution. Yet despite the long history of technology assessment, we know that there is a major difference in the present situation: somehow the available assessment institutions and mechanisms are inadequate to the tasks awaiting assessment today; and we suspect that they will be even more deficient for future assessment challenges coming over the horizon. In short, we know how important it has become for society to improve its process of technology assessment.

Second, there appears to be a consensus regarding the pervasiveness of the assessment process and the consequent requirement for

*Presented May 26, 1970.

a pluralistic approach toward its improvement. We know that Congress plays a crucial part in major public assessments, but we recognize as well the essential roles played by the Executive Branch, the Judicial System, state and local governments, industry, the nonprofit sector (including universities), citizens groups, and the public-at-large. While particular assessment activities may temporarily place one or more of these "assessment arenas" in the forefront of decision-making, there is no fixed hierarchy of importance which can be ascribed to these assessment arenas for all, or even most, classes of assessment decisions. Choices made by the Congress or the White House may be fraught with significance, but so might be key decisions of industrial leaders, the total impact of millions of individual consumer choices, and the cumulative effect of innumerable, impartial, judicial determinations.

Given this pervasiveness of the assessment process, together with the primitive state of our knowledge regarding its workings, the seminars yielded a consensus concerning the need for a pluralistic, experimental approach aimed at ameliorating the process of assessment. Thus it was generally recognized that there was no panacea possible in this field, but that all assessment arenas would have to be significantly enhanced and woven together into a viable assessment system, which would accord adequate consideration of all relevant factors. In view of our sizable ignorance in this area, many of the seminar participants pointed out the need for extensive experimentation in the evolution of a more adequate assessment system.

Third, there appeared to be a general consensus regarding the desirability of performing "total systems assessment," insofar as possible. The emphasis here was on taking balanced account of the following: 1. the positive as well as the negative consequences of technology and 2. the secondary, tertiary, and higher-order consequences, as well as the direct impacts of technology. Moreover, the context of analysis has to include the full range of social, economic, political, legal, and psychological factors which impinge on technology. While it was recognized that the state-of-the-art regarding these factors left a great deal to be desired, it was generally agreed that we had to press ahead on these fronts as effectively as possible.

It was also pointed out in the seminars, however, that we could not expect a "total systems assessment" to be followed by all analysts throughout the assessment structure. Many, if not most, assessment analysts were constrained by their limited set of organizational objectives and responsibilities to take a more parochial view of any particular assessment problem. The practical goal, therefore, was not adherence to total system assessment by all participants in the assessment process, but rather that the total systems approach be followed by at least some sufficiently prominent participants in the process, whose results would be adequately considered by the key decision makers with respect to that particular assessment.

Fourth, there was a general consensus that there was a need for much better, more comprehensive information provided to Congress to enable it to make truly informed judgments regarding technology assessment. It was pointed out, of course, that the quantity of information in existence was not the problem. Frequently, in fact, the amount of information in existence far exceeded the capacity of Congressmen and their staffs to absorb it. The problem arose rather from the quality of the information and its pattern of presentation and incorporation into the system. The deficiencies in quality developed because total systems assessment was only rarely employed. The problems of presentation and incorporation stemmed from the complexities of Congressional committee structure and the gerrymandered jurisdictional labyrinth within which issues have to seek clarification and comprehensive consideration, as well as from the unremitting pressures of special interest groups seeking to promote their own points of view through incomplete or biased assessments. (It may be noted in passing that although there appeared to be considerable consensus concerning Congress' need for better information, there was little, if any, agreement regarding the ways in which that need should be met.)

Fifth and finally, there appeared to be some measure of consensus that adversary proceedings comprised a significant aspect of the assessment process, even if the precise role and significance remained under dispute. Some seminar participants asserted that adversary proceedings--within the legal system, between competing economic interests, between contending Congressional pressure groups, between various Government agencies--constituted the ultimate forum in which assessment decisions were reached. Others maintained that this pattern was not only prevalent, but was desirable and healthy as well, while still others believed we should strive to reduce the adversary aspect of assessment to the minimum extent possible.

NATIONAL ASSESSMENT SYSTEM

Turning from these points of consensus to the expression of my own views, I shall first sketch out my version of a national assessment system in order to provide necessary background for the examination of citizen participation.

The system I envisage would encompass all assessment arenas: the three branches of the Federal Government, state and local governments, industry, the nonprofit sector (including universities), citizens groups, and the public-at-large. Below are brief statements of the structure and role of each of these components of the assessment process.

Each Federal Executive Branch agency to which technology assessment is relevant should carry out its assessment function through an office or an individual reporting to the head of the agency. This does not mean that there might not be component or subordinate assessments, which would be carried out further down the organizational structure in the agency or department. But it does mean that, notwithstanding any assessment at lower levels, it is essential that a technology assessment activity be located at the very top of the agency so that it can independently assess the technologies which may be advocated by more narrowly oriented components of the agency.

In addition to assessment at the agency level, there must be assessment of significant matters at the White House level. Most discussion of assessment in the Executive Offices of the President has focused on the relative roles of Office of Science and Technology (OST) and Bureau of the Budget (BOB), with some consideration given to the Environmental Quality Council. However, under the President's Proposed Reorganization Plan (which will presumably take effect shortly, barring a last minute veto by the Senate), I believe the following assessment organization to be more appropriate.

The Domestic Policy Council would have responsibility for significant technology assessments in the Executive Branch. According to the Reorganization Plan, this Council would consider what programs, in what priorities, the Government should mount. The Office of Management (in which the former BOB would be incorporated) would, on the other hand, consider how best to carry out the various programs and would appraise their subsequent performance. In view of these relative roles it appears desirable for the Council to have the primary assessment responsibility, receiving staff support from the Office of Management, the Environmental Quality Council, and the OST.

Assessment in the Congress is, of course, the subject of the pending Daddario Bill, on which hearings are currently being held. It is clear that Congress does need augmented assessment assistance, although one can dispute the organizational locus for that assistance, with powerful pros and cons associated with every alternative (i. e., Library of Congress, Government Accounting Office, an independent

Office of Technology Assessment). Whatever the outcome of the current bill, Congress should certainly be provided with a much strengthened technology assessment unit, which could initiate studies through its own staff or by means of contracts with outside research organizations.

In addition to strengthening its in-house staff resources, Congress should charter an independent research organization, The National Institute of Technology Assessment (NITA). Congress should provide NITA with a long-term endowment, supplemented through the appropriations process. NITA should also be empowered to seek private sources of funding as available. NITA would provide a continuing, independent, prestigious organization which could carry out in-depth, total systems approach technology assessment. The input of such assessments into the Congressional process would provide an element of independent objectivity and depth of analysis which would probably not be attained through the proposed Office of Technology Assessment alone. NITA could, in fact, be a major contractor of such an Office.

While it is far beyond the range of this paper to delve into assessment at the state and local levels, it is clear that all tiers of government require assessment capabilities appropriate to the scope of their responsibilities.

The nonprofit sector of society, including the academic community, must play an important part in any effective system. Despite its resources, NITA could handle only a limited number of the issues and areas requiring technology assessment analysis. Technology is far too pervasive in our society for one major research organization, such as NITA, to perform the wide range of research which is required. Full participation from the nonprofit sector and academic community will be necessary for a successful national assessment system.

Industry has always played a crucial role in the assessment process and will undoubtedly continue to do so. The only point to be emphasized in this connection is that industry would be well-advised to expand the time scale and range of factors it takes into consideration in its assessment decisions. As political action and judicial determination increasingly internalize social costs in industry, firms will find it more to their own long-term advantage to attempt to incorporate social impact thinking in their current assessment decisions.

The Judicial System has always been, and will undoubtedly continue to be, crucial to the evolution and operation of technology assessment processes. As has been pointed out by Milton Katz in his testimony before the Daddario Subcommittee

I would like to emphasize one point about the assessors, who are in fact the whole business community. Whether or not they are aware of the fact, the existing legal order infuses their calculations. It is the legal order that determines which of the anticipated costs and benefits are taken into account by the enterprise and which are ignored. An electric light company, for example, which contemplates the installation of a new power plant will treat the fuel to be consumed as a cost, but not the smoke that may pollute the surrounding air, nor any waste products that may be discharged into nearby streams. The company's management may anticipate a public relations problem from the pollution. If you look hard at what they mean by a public relations problem, it appears that they have in mind a risk that an aroused public opinion may generate changes in some applicable aspect of the legal system.

In the ordinary course of business, the company will calculate the estimated costs and benefits of the prospective installation without reference to any damage to the community caused by the smoke or other waste products. In the language of the economists, pollution of the community's air or streams through the operation of the power plant would be a "social cost," not a cost of the enterprise; it would be an "external" not an "internal" cost.

The economic mode of analysis is an indispensable tool for technology assessment. But I want to emphasize that the economic analysis takes for granted a particular posture of the existing legal system. Why is damage to the community caused by waste products from our hypothetical electric powerplant a "social" cost or an "external" cost? It is a "social" and "external" cost only if and to the extent that the legal system happens so to decree. The legal system can alter the incidence of a cost by recognizing a cause of action in tort against the company. A judgment in tort will transfer back to the company the cost previously suffered by the community in the form of air or water pollution.

The judgment then converts what the economists call a "social" cost into what the economists call an "enterprise" cost. It internalizes the so-called external cost. In a similar way the legal system can maintain the incidence of a cost by declining to recognize a cause of action in tort against the company.

Through tort law, the legal system operates directly upon the incidence of costs. Through the law of contract, the legal system may operate indirectly upon the incidence of costs. Contract law may enable the persons involved to adjust or modify the incidence of a cost by giving effect to agreements among them designed to effect such an adjustment. On the other hand, contract law may frustrate efforts of the persons involved to modify the incidence of costs by declining to give effect to agreements among them designed for such a purpose.

In the long history of the common law in America, changes have occurred from time to time affecting the incidence of costs. Changes have also been made by legislation, such as industrial safety and accident legislation and workmen's compensation laws. Comparable changes may occur in the future in the continuing evolution of the law in response to the changing realities of American life.

Let me take a moment to hammer the point home. When is it a good business proposition to put something on the market? From the point of view of the business enterprise, it is a sound step if the product to be marketed will make money. I want to emphasize first, that the enterprise's own estimate of anticipated income and expense takes for granted the existing provisions of the legal system; second, that the existing provisions of the legal system at any time are the result of a long evolution; and third, that the legal system in America continues to evolve. Any changes which you gentlemen may make through new legislation will become part of this continuing evolution. You would not be altering the basic structure of the legal system nor of the business system. You would be altering the incidence of costs whose incidence has been altered before in the evolution of the business and legal systems.[1]

[The final element in the nation's assessment system is the essential role of citizen participation, the topic to which the remainder of this paper is devoted.]

THE NEED FOR CITIZEN ACTION

It is believed that the national assessment system sketched out above would afford a feasible framework of institutions and mechanisms, within which the assessment process could proceed effectively, provided that the essential element of citizen participation is forthcoming to the extent necessary for assessment to reflect the underlying needs and demands of the society. Technology is interwoven throughout the fabric of our society; yet as widespread as is its role today, its potential range of ramifications is likely to be even more extensive tomorrow. As war is said to be too important to be left to the generals, so technology assessment is far too crucial to the shape of our future to be left to the professional assessors and the special interest groups involved, regardless of how excellent their qualifications or how altruistic their objectives may be.

The world of tomorrow will be increasingly a technological society. Technology assessment, regardless of how recondite its details may be, must become an integral aspect of the nation's total social, political, economic decision-making processes, in which all citizens have the opportunity to participate. Otherwise, in a technology-permeated society, it will become increasingly difficult--if not impossible--to maintain, much less enhance, the democratic character of our society and the quality of freedom in our lives.

Hence, citizen participation must be an absolutely essential aspect of the assessment process. There are innumerable impediments, however, which stand in the way of citizens' taking effective assessment action. These impediments fall in three interrelated areas: 1. finance, 2. organization and motivation, and 3. information, each of which is treated in turn below.

1. Finance

With rare exceptions, individual citizens do not have the financial resources to enter deeply into the assessment process as individuals. Those who are professional experts in a particular area can often express their assessment views in the open literature, in Congressional testimony, or as expert consultants. But apart from this group (which is generally quite small relative to any particular issue), and from the even smaller group of extremely wealthy individuals who make a practice of espousing causes, individuals singly cannot play a significant role in the assessment process, except through exercising their power of choice in the operation of the market or political system.

By joining together in groups, citizens can, of course, exert a much greater influence, depending on the size of the group, its financial resources, and its cohesiveness with respect to the issues under contention. We are all familiar with the diverse conservation, environment, and consumer groups which have been proliferating in recent years, not to mention the various political action groups which have been emerging recently in response to issues such as the Viet Nam War, civil rights, etc.

One problem common to almost all such groups is inadequate financing; fund raising is usually a persistent problem, and much of these groups' energy and effort is generally devoted to replenishing their depleted coffers. This lack of money imposes severe limitations on the influence such groups can exert, especially vis-a-vis well-financed special interest groups with which they may be contending, either for broad public support or for Congressional decisions.

Effective citizen participation in the assessment process requires new financial mechanisms whereby such groups can obtain the necessary funds, on a continuing basis, to compete on an even-footing with the well-heeled special interests. Below, in this paper, is presented a specific proposal for meeting this financial problem.

2. Organization and Motivation

The problems of launching such groups, of organizing them for effective action, and of motivating citizens to join them and to support their efforts are intimately intertwined with one another and are all dependent on meeting the essential financial prerequisites. For example, consider the problem of motivation. This problem does not apply to the initial formation of the group by a small number of highly motivated individuals, but rather to the difficulties involved in motivating large numbers of members to join and actively support the group's programs. Assuming that the group addresses a real need in our society and has some inherent appeal for some segment of the public, then the problem becomes one of proper promotion of the group's objectives and programs and the values associated with membership. This in turn resolves into a financial problem: if sufficient seed money is available, then an effective promotional campaign can be mounted and additional members obtained, who in turn generate additional funds.

The problem of organization is similarly dependent on financial considerations. Many such groups are reported to be relatively inept at developing a strong internal organization and at structuring their external relationships with Executive agencies, the Congress, the public-at-large, or other specific groups they may wish to influence. But I suspect that whatever ineptness there may be, in fact, is probably due far more to limitations in funds than to any lack of potential for the pragmatic exercise of power. One may certainly cite examples of highly successful efforts supported with meagre financial resources. (Ralph Nader, of course, started out by himself; and even today I doubt whether the powerful "machine" he leads is exactly affluent, especially vis-a-vis the interests he and his adherents are opposing.) But the fundamental point, I think, remains valid: that given a group with inherent appeal to some segment of the public, the problems of motivation and organization are largely dependent on the financial resources which can be obtained.

3. Information

While the problem of obtaining and utilizing information effectively is also dependent in large part on the availability of adequate financial resources, there are also research problems and time delays involved here which are of crucial importance. For example, one may know that the dumping of industrial wastes into bodies of water has deleterious consequences without knowing the nature and extent of those consequences or the relative damage contributed by particular components of the overall mix of industrial wastes. Answers to such questions, however, are frequently extremely important to the design and implementation of practicable anti-pollution programs. Yet obtaining valid answers often requires extensive research, and the research in turn entails time and money.

The difficulty of obtaining adequate information is further complicated by the fact that many of the consequences of technology, beneficial as well as adverse, do not occur--at least sufficiently so that they can be identified--until considerable time has elapsed, and vast resources have already been irretrievably committed to particular courses of action. Once such resources have been so committed, powerful special interest groups are generated with the objective of maintaining and enhancing their stake in the technology under consideration.

In view of this situation, the performance of adequate technology assessment entails the incorporation of an "early warning" capability which can identify such problems well before they arise and before the related patterns of resource allocation have been cemented in place. The development of such an early warning requires a great deal of additional research and experimentation in such areas as technological forecasting, social indicators, and the application of systems analysis to social and behavioral problems. Society still has a long way to go in devising appropriate early warning techniques.

In addition, there is a corollary capability which must be developed if early warning efforts are to prove of any avail. This is the capability to take appropriate action, after society has been duly forewarned. In certain limited areas, there are some existing mechanisms directed at this objective. For example, if a food additive is shown to induce cancer in a test animal, its use is prohibited. Similarly, if tests of new drugs show them up as ineffective or as yielding adverse consequences which outweigh their positive effects, the drugs can be prohibited. Thus in a few areas, there are mechanisms, however imperfect they may be, for implementing the results of early warning research. In the vast majority of instances, however, in which technology can impinge on society and human life, there are no adequate mechanisms for acting upon early warning results. Thus, for example, if a new type of container material were developed today which research indicated would come to constitute a serious environmental pollutant two decades from now--after certain quantities had accumulated and certain chemical changes had occured--there would be no way of halting production, short of new legislation specifically aimed at that product.

Another approach to the problem would be general legislation calling for a vast expansion in government regulatory control over industrial operations and products. This would be bound to have an inhibiting effect on the rate of technological innovation and would probably dampen the overall vigor of the economy. At the same time it would go a long way toward radically altering the balance of power in the country between the Federal Government and private enterprise. In any event, it is not an approach likely to achieve widespread support and enactment in the foreseeable future.

The fundamental problem remains, however, for technology assessment to prove effective, society must have the research capability to perform the early warning function as well as the implementation mechanisms whereby such warnings can be acted upon before it is too late.

We can summarize the requirements for effective citizen action in the assessment process as follows: Society must afford existing and prospective citizens groups the opportunity to obtain adequate financing on a continuing basis. With such financing, citizens groups can motivate their potential membership to join and participate, and can organize themselves for effective action. They can also use the financial resources to obtain the necessary base of information to further their causes, supporting the performance of research when necessary. Furthermore, as the early warning capability is perfected, they can assess the future consequences of current and projected technologies. Finally, as mechanisms are developed whereby society can take prompt action in response to the results of early warning research, citizens groups can come to exert the extensive influence they deserve to wield in shaping the course of the future.

CITIZENS' ASSESSMENT ASSOCIATIONS: A PROPOSAL FOR ACTION

General

The following proposal has been designed to meet the objectives outlined above. It does not purport to be a finished end product, but is put forth as a preliminary proposal solely to serve as the basis for further thought and discussion along these lines.

The proposal calls for the establishment of citizens' assessment associations (caa's) whose functioning would be fostered and regulated by a new Federal agency, the Citizens' Assessment Administration (CAA). Through the financial mechanisms described below, the caa's would be enabled to obtain adequate funding on a continuing basis, which would provide them with the essential financial resources required to assert significant influence in the assessment process. With this financial base, they would be able to promote their objectives and activities, motivate sufficient numbers of their potential membership group, and organize themselves for the effective exercise of influence on the assessment process. To cope with the important information requirements for effective assessment action, they would be empowered to assemble, process, and analyze information relevant to their assessment topics and whenever necessary, to conduct or commission necessary research relevant to their assessment areas.

When caa's had accumulated and analyzed relevant information needed to perform the desired assessments, they would be empowered to disseminate the results of their assessments to the public-at-large, as well as to appropriate decision-making organizations within the society (Congressional committees, Executive agencies, etc.). They would thereby perform a public information function, as well as be in a position to lobby for legislation or executive regulations in keeping with their findings.

In addition, however, they would have the extremely important power to institute legal, class action proceedings against any organization or individual within the society (including agencies of Federal, state, and local government), which were making use, or planning to make use, of technologies whose assessments indicated detrimental consequences to the persons or interests of certain segments of the public. These functions of the caa's, along with their facilitating mechanisms, are discussed in turn below.

Citizens' Assessment Administration

This would be an independent government agency with its Administrator reporting directly to the President. The Administrator would formulate and carry out the policies of the agency within broad guidelines laid down by a Citizens' Assessment Board, whose members would be appointed by the President, and who would represent a wide spectrum of interests in American society.

The CAA would be responsible for developing criteria for, and regulating the establishment and functioning of, citizens' assessment associations. In addition, the CAA would administer various financial measures (described in the section on Financing below), which would be designed to protect the viability of citizens' assessment associations. (Although there are many substantial differences, the relationship of the Small Business Administration to small business firms can be thought of as somewhat analogous to what is intended here.)

Establishment and Organization of caa's

Any group of citizens, meeting the criteria set forth by the CAA, could establish a new citizens' assessment association. In addition, existing nonprofit organizations could be converted into caa's, if they met the necessary criteria. The purposes for which a particular caa could be formed would be as broad as "protection of the environment" or as narrow as "assessment of consumer products containing asbestos." The specific purposes would be spelled out in the caa's incorporation charter within guidelines established by the CAA. The initial financial support for caa's could come partly through individual donations and membership dues and partly through foundation grants or Government grants and contracts. In addition to these currently available sources of funds, caa's would also have the new mechanism available of issuing citizens' assessment bonds (described below). These bonds would provide caa's with the continuing financial stability essential to making a real impact on the assessment process. Once established, the new caa would be empowered to use a portion of its funds for promotional purposes to sell more citizens' assessment bonds and to increase its membership. There could be different classes of membership and voting rights depending on whether an individual or affiliated organization made a contribution, paid dues, or purchased a bond. (The CAA would have to regulate these matters carefully to preclude the seizure of control of a caa by contending economic interests, e.g., the purchase of a controlling amount of bonds in a caa oriented against oil spill pollution by the oil industry.)

Functioning of caa's

The primary purpose of each caa would be to perform technology assessments in its areas of interest or to draw upon assessment results obtained by others, and to utilize those results to affect the decision processes regulating society's use of the technology or technology-based system under consideration.

To accomplish that purpose, each caa would have in-house, or available to draw upon, a capability for arriving at assessment judgments. Thus the caa could have its own staff and/or advisory council of assessment authorities, which would form the assessment judgment upon which the caa would act. Or the caa could draw upon available results of assessments by groups such as the National Academies of Sciences and Engineering; or contract with universities or research institutes, to carry out specific assessment assignments. When further research was required before an assessment judgment could be formed, the caa could similarly carry out such research in-house or contract with others for its performance.

Regardless of which of these patterns was followed, the caa would arrive at an assessment judgment upon which it wished to act. (Since the essence of the caa concept is citizen participation, there should be provision in the agency rules regulating caa's that such assessment decisions must be duly ratified by the caa's membership before they can be accepted and acted upon. This would help preclude the caa's from being subverted into elitist, expert-dominated organizations.)

Upon acceptance of an assessment, the caa could follow one or more of a number of possible courses of action:

1. The caa could disseminate its results publicly and attempt to influence overall public opinion or the views of selected segments of the public.

2. The caa could lobby directly (or indirectly through other lobby organizations) with Congress, state legislators, Government agencies at, Federal, state, and local levels, the White House, governors, influential private organizations and individuals, etc. The purpose of such lobbying would, of course, be to induce the target group to accept the assessment results and take appropriate action on them.

3 The caa could institute class action and legal proceedings on behalf of its membership and other potentially affected parties. These

legal proceedings could be directed at any organization or individual in society (including agencies of Federal, state, and local governments when appropriate), which were making use--or planning to make use--of technologies whose assessments indicated detrimental consequences to the persons or interests of certain segments of the public.

This power to initiate litigation includes several important components:

1. The suits would be class action suits that would apply to whole classes of affected parties.

2. The detrimental consequences could be either to the "persons" or the "interests" of certain segments of the public. Thus if it could be shown in court that it was to the interest of a certain segment of the public to maintain the beauty of a national park intact and uncontaminated, then action would injure that park would be detrimental to the interests of the affected segment of the public.

3. The technologies with the detrimental consequences need not be functioning already for the litigation to commerce. The planned use of technologies with detrimental consequences would also be subject to appropriate litigation.

What kinds of results would ensue from such litigation? In the case of technologies which were already in operation, with attendant detrimental consequences, the courts could award damages to the CAA and associated affected parties. (In addition, appropriate criminal action could be initiated when criminal violations had occured.)

In determining damage awards, the CAA Act establishing the agency and the associations would extend the concept of damages and associated costs to include not only real and punitive damages when applicable, and the litigation costs borne by the caa, but also that portion of the caa's operating costs which enabled it to prosecute the suit successfully. Thus the caa would be entitled to be reimbursed for: 1. its own operating costs relative to the preparation for and prosecution of the suit; 2. the costs of relevant research contracts and consulting fees; and 3. an appropriately prorated portion of the interest on the caa's citizens' assessment bonds. This statutory extension in the concept of damages and associated costs would go a long way toward assuring the financial viability of caa's.

In the case of technologies whose detrimental consequences had not yet occurred, the following kinds of results would be possible.

This would include technologies which were planned but not yet in being, as well as technologies in existence, whose detrimental consequences had not yet occurred, but could be scientifically forecast with some degree of confidence.

In such cases, the caa could seek permanent injunction to prohibit further implementation of the particular technology, as well as appropriate dismantling of what was already in being. If such an injunction were awarded (and sustained of course), the organization on whom the injunction were placed would be liable to reimburse the caa for its litigation costs, and also for the associated costs necessary to prepare the case (as outlined above). Again this would greatly aid the caa's maintaining financial viability.

Financing of caa's

As noted above, caa's would be permitted to accept charitable donations, membership dues, and grants and contracts from private and government organizations. But the primary source of their funds, and the foundation of their financial stability would be citizens' assessment bonds they would be empowered to issue.

The interest rate on these bonds would be regulated by the CAA. The rate would be set at a higher level than that permitted on savings bank accounts, and probably somewhat higher than that permitted on bank certificates of deposit or savings and loan association rates.

Bonds would be issued for ten year periods, and interest on them would be guaranteed by the CAA in case of default on the part of a particular caa.

Other sources of funds available to caa's, besides the bonds, donations, dues, grants, and contracts, would be reimbursement for operating costs (as broadly defined above) arising from successful litigation. caa's would redeem outstanding bonds at the end of ten year periods through these sources of funds, as well as through additional bond issues.

With these ground rules, some caa's would undoubtedly become financially insolvent. In such cases, the remaining assets of the caa would be distributed to the bond holders on a *pro rata* basis. Thus there would be some element of risk in these bonds; and it is for this reason that their interest rate would be set somewhat higher than bank savings certificates, for example.

The interest rates would not be set too high, however, because the purpose of these bonds is not to provide a desirable form to investment in general. Its purpose is instead to enable citizens who care about particular assessment issues, like water pollution or noise due to the SST, to contribute to society's resolution of the issue, at minimal risk to their normal savings.

To those who would doubt the appeal of such bonds, I would merely point out how voluntary citizens organizations have managed to survive financially without this reimbursement mechanism. With it, I think they will flourish, and citizen participation will rightly become a powerful factor in the assessment process.

Balanced Approach of caa's

Although much of the preceding discussion on the CAA has been couched in terms of the negative consequences of technology, there is nothing inherent in the caa concept to exclude the promotion by caa's of particular technologies with expected positive consequences. For example, a caa could be formed to promote the development and use of electric cars, or certain systems of public transportation, or solar energy systems, etc. The purpose of the caa concept is not to facilitate citizens' attacks on technology, but rather to enable citizens to achieve full democratic participation in the process of technology assessment. Without such participation, the whole character and quality of our democracy would ultimately be vitiated.

IMPLICATIONS OF CITIZEN ASSESSMENT ACTION

The citizens' assessment association concept as presented represents an institutional innovation which could prove significant. Accordingly, it is worth exploring some of its major implications.

If the caa concept were implemented, there would probably be extensive use of class action suits. At the same time the proposed legislation would foster an extremely broad interpretation of the interests of certain segments of the public. Recent lawsuits filed in the environmental area have been filed partially on behalf of future generations. While this is perhaps an extreme case of a broad interpretation of 'interests', it is nonetheless the general direction toward which the caa concept would move litigation.

The broad interpretation of associated costs of litigation, to include the operating and research costs of the caa necessary to establish the assessment case, appears to be of some legal significance. Whatever its significance, however, I believe this interpretation is absolutely essential to enable the judicial system to play the crucial role with respect to society's utilization of technology that it has played in other areas of society's evolution. For the complexities of technology are so great and the future consequences of present technological activity are so difficult to determine, that relevant research must be seen as an essential aspect of litigation on these matters.

Finally, there are the implications of using present scientific research as evidence to assert that it is reasonable to conclude that certain consequences are probable to occur in the future. For example, scientific research could well conclude that the use of a certain chemical in small doses over a period of time would cumulatively constitute a future hazard to a statistically significant segment of the population. On such grounds under the caa proposal, an injunction could be obtained against the promoter of the chemical, with his incurring a financial liability to the caa which successfully sought the injunction. This seems to me again to pose some legal issues of apparent significance. But I am convinced that the legal system must find a way of taking account of such considerations, if the system is to fulfill its role in a technology-centered, highly interdependent society. Put in other terms, I believe the legal system must find a way of making present determinations of fact on the basis of scientific evidence regarding probabilities of future occurrence.

As challenging as some of these problems may be to the legal community, the industrial world will also have its share of adjustments to make. The concept of imposing costs on an industrial firm on the basis of some of its planned activities, or on the future consequences of present activities which are scientifically forecast to prove detrimental, is certainly something of a radical notion. But again, I am personally convinced something of that sort is essential for society in order to control the evolution of technology-based industry in socially desirable directions-- or at least in directions which are not socially detrimental. One point is clear in this regard: if such costs were imposed on industry, firms would certainly think much more deeply and carefully with regard to the social consequences of their actions. The "total systems, future-oriented approach" to technology assessment would undoubtedly gain many adherents in industry.

A final implication of the caa concept which may be cited derives from the citizens' assessment bonds. These bonds are predicated on the assumption that it is proper for citizens to receive a financial return (even if a modest, limited one) on relatively low-risk investments they may make, with the objective of enhancing the overall assets of society, such as environmental quality. It is interesting to ponder where such a radical notion may eventually lead.

CONCLUSION

In this paper I have made a plea for the importance of citizen participation in the assessment process and presented a proposal for an institutional innovation which would facilitate effective citizen assessment action. As I stated initially, the proposal is a preliminary one intended to generate discussion on the myriad of issues involved. It contains a number of radical concepts and mechanisms, which are undoubtedly open to a number of criticisms. Nevertheless, I believe the proposal contains the germ of an idea which is worth pursuing. If recent decades have taught us any lesson, it is that the radical concepts of one year rapidly become the cliches of the next one. On one final point I am absolutely convinced: we have to find a way of assuring effective citizen action in the assessment process if our society is to survive as a democracy in which the quality of individual life remains paramount.

NOTE

1. U.S. House, Committee on Science and Astronautics, Subcommittee on Science, Research, and Development, Technology Assessment, Statement of Milton Katz (December 4, 1969), Committee Print, 91st Cong., 1st sess., pp. 174-75.

DISCUSSION OF TECHNOLOGY ASSESSMENT AND CITIZEN ACTION

In reviewing his paper, Mr. Mottur noted that "although citizen action is very important, I don't by any means want to imply that, by itself, it is a panacea" for the problems of our technological society. All segments of our society must be involved in the technology assessment process; citizen's groups are but one part of the society, albeit, in this context, a very important one.

Why is citizen participation important? Mr. Mottur admitted that "my reasons for believing that citizen participation is important are emotional and based on a certain conception, which I have, of what American democracy is." The continued predominance of "the elitist decision-making system which. . . is too prevalent in the science and technology area of our society" can be avoided only through the development of a viable means of citizen participation in important decisions involving the impact of science and technology upon society.

Perhaps the major impediment to effective citizen action is the difficulty of obtaining adequate funds to "compete with special interests. . . . in that adversary process which is the. . .guts of the assessment process." The funding scheme outlined in the paper, Mr. Mottur emphasized,

> . . . is not a mechanism whereby citizens can make a lot of money out of advocating social progress. It is a mechanism whereby citizens who care about social progress--or who can be induced to care about it--can feel that they are making an investment in advancing certain social goals.

One of the novel ideas incorporated in his proposal, Mr. Mottur felt, was the notion that an industrial firm against which an injunction was granted would be required to contribute to the costs of the citizens' assessment association which argued for the injunction. An important general principle is that "since technology is such that if a business firm makes a decision today, it may be ten or twenty years before its effects are felt," industry must be held responsible, at least financially, for determining the potential adverse effects of its actions.

But the caa's and technology assessment itself, need not be totally negative in outlook. "Although I honestly believe that in order for these groups to get started much or the initial emphasis would be on the negative impacts" of technology, over the long haul the caa's "could push the positive side of technology as well." As an example he cited the possibility of a caa organized around the promotion of electric cars as an alternative to gasoline driven vehicles.

Mr. Mottur concluded his summary by noting that:

> . . . it's becoming increasingly difficult for a democratic society to control. . . complex technological systems and to do it through the decentralized, socially responsive decision-making that is at the heart of the democratic system. I think that unless we get a broadly conceived form

of technology assessment going in our society, there is not much hope for the survival of our civilization. But even if we get it going, it will not be working properly if it is not deeply rooted in widespread citizen participation.

THE PROPOSAL FOR CITIZENS' ACTION GROUPS

A part of the discussion centered on certain aspects of Mr. Mottur's specific proposal of a mechanism for citizen participation in the assessment process. A lawyer felt that the proposal certainly "highlighted the critical problem: financing." Another participant noted the great difficulty which citizens' groups which protest aircraft noise have in obtaining funding which will allow them to compete with the strong and well-financed interests of the aviation industry. Mr. Mottur called attention to the fact that "public interest" efforts have, in recent years, been funded by private foundations. In his view, however, foundation funding was not an adequate solution to the long-term funding problem. Changes in the tax structure and a decrease in the "relative portion of available wealth in the society that can come from these private foundation," will, in his view, make the problem of funding even more critical in the coming years.

A university professor was somewhat skeptical of the ability or inclination of citizens' action groups to become involved in a broad range of problems. In his opinion, the caa's of the proposal would "only function in those areas where they can smell, see, or hear something they don't like. I think groups of this sort are already functioning well in those areas." But Mr. Mottur disagreed. He felt that many problems beyond those which can be easily sensed could become subjects of attention for citizens' groups if only a "small nucleus group. . . can get enough others interested in doing something about the problems." The average person today feels impotent, but "something like this [a caa] would provide a mechanism for people to influence decisions" and so make people more involved in a broad range of policy issues.

Another professor felt that any citizens' action group would have its own special interest, its own position on a given matter. If this is so, then why is there a need for "going through the motions of a technology assessment?" "Why," he asked, "should the group waste its time, money, and resources on doing assessments," if what it really wants, no matter what an assessment shows, is that its own position should be upheld. Mr. Mottur felt that in order for the group to make "specific, pragmatic proposals that would affect the policy-making process" it would have to perform, or have access to, a technology assessment of the problem under consideration. This is not to say that the caa's must always perform assessments; "there is a large inventory of items on which assessments have been done so that these groups could, in some cases, use the existing assessments and move from there."

A Government official saw a potential difficulty in that aspect of the proposal which dealt with foreseeing the future effects of technology:

> You can prove that some technology which has been in use has had some adverse effect. . .; it is impossible to prove that a prospective technology will not turn out in the end to have had some adverse effects. And if you make the burden of proof upon the introduction of new technology too great, what you will do is bring innovative activity of any kind to a crunching halt.

Mr. Mottur agreed with this point and felt that "very strong safeguards" would be necessary in any system that was implemented. While he felt that industry must bear some burden of the costs of examining the potential consequences of its actions, society should also carry some of the costs in order to assure that innovation will not be completely stifled.

But there were those who felt that the inhibition of innovation was precisely what was called for in order to minimize adverse effects of technology. One participant held that, at least in some areas, such as in some of the new biomedical technologies, innovation might best be prohibited. He continued:

> It seems to me that one issue that a process of assessment should focus on is an articulation of the values which are served by a particular innovation. Even if you had adequate criteria for determining acceptable levels of risk and if you determined that a given technology was sufficiently safe, you should still ask the question, "Who needs the technology?" Why should we allow the spending of the public resources for this particular technology? I think we have to get away from the very nineteenth century notion that all change is progress and all progress is good.

THE ROLE OF EXPERTS

Mr. Mottur noted that "citizens' groups would draw on experts as would anyone else." Two participants noted that experts were in limited supply, and Mr. Mottur agreed that this was a real problem for citizens' groups. Another participant questioned whether the information that citizens would get from experts would be any different from that which is obtained now. Mr. Mottur was emphatic in contending that there would be a major difference:

> Information on technology assessment matters can almost never be divorced from value considerations. The assumptions that are made in determining what the so-called facts are in looking at any system are really quite critical. While obviously the citizens' groups will have to rely on many of the same experts that other people will be relying on [because of the limited supply of experts mentioned earlier], the manner in which they frame the questions and structure the investigation as well as how they integrate and present the results is of great importance.

A university professor was concerned about a general attitude toward experts. "Having decisions made by experts has been equated with rationality, innovation, and progress" while having nonexperts involved leads to talk of irrationality, stagnation, and destruction. He held that if you rely on experts, "you are never going to have the right questions asked or the right questions answered" because experts tend to ignore things which are readily perceived by the most unsophisticated nonexpert. Pursuing a related line of thought, another participant pointed out that experts often overstep the bounds of their expertise and become involved in matters on which they are no more qualified to make judgments than anyone else. In discussing hospital committees which make decisions on human experimentation he noted:

> These committees are almost universally made up only of physicians. I would suggest that this is an intolerable situation because the issue of whether a particular experiment on a human being is worth doing or will be done with proper attention to certain human values . . . is simply not an issue on which the doctor has any better information than anybody else. Indeed, the doctors'

> professional expertise is normally going to get in the way of a decision that is acceptable to the society or to the patient.

A lawyer suggested that the proper role for scientific expertise in citizens' groups is not in actually making decisions but in assisting the groups and their lawyers in "marshalling evidence as effectively as possible."

THE ADVERSARY SYSTEM AND TECHNOLOGY ASSESSMENT

The role of the legal system and the adversary process was discussed at length during the session because of the central place which the adversary concept had in Mr. Mottur's proposal.

A Government official was wary of placing too much faith in the ability of the legal system to assist in arriving at decisions in the public interest; certainly in the area of food additives and new drugs, he claimed, "the legal system is being used now to protect those who are abusing the masses of our citizens." Mr. Mottur responded that the law is a changing thing; it is changed both by statute and by continuing "evolution in the way in which laws are interpreted." By fostering changes in the legal system we can make the law more responsive to the needs of the citizens.

A Congressional Aide was very concerned by the almost insurmountable problems for the courts which are caused by the "many extraordinary expertise problems" involved in technology assessment and the large component of morality and value judgments which must be taken into account. Another participant agreed that the problems were enormous but noted that the courts could not avoid making decisions in this area "without recognizing that in not deciding they are, in fact, making a very important decision."

A Government official questioned the efficacy of the adversary process as a mechanism for performing technology assessment. He felt that the assumption that the adversary process would be helpful in this regard was made without proper examination of its strengths and weaknesses. The lawyers present rose to the challenge. One participant held that:

> As a basic information device the adversary system has never been beaten. I don't think this has anything to do

> with the private bar. If you go back into any society, no matter how primitive, you will find that they always try to get at both sides of a question because they simply find that they can't really figure out what is going on otherwise.

A university professor had this to say:

> We are living in a society in which we find the adversary process at the heart of both the economic and political system an adversary process which pits buyer against seller in the market, an adversary process in which our entire political form turns upon decision made when adversaries argue in election campaigns or in our legislative bodies. As Oliver Wendell Holmes said many years ago. "There is no better test of the truth of an idea than its acceptance in the marketplace of ideas where conflicting views clash."

Still another participant stressed the pervasiveness of the adversary system: "Society, more or less throughout history, has been characterized by some kind of third party decision-making process where people take positions that are resolved by others than the proponents of the positions." But a practicing attorney found the adversary process somewhat less pervasive in our society than the previous speakers:

> It seems to me that the essential problem with the adversary process is that as far as issues of technology assessment are concerned, the adversary process has never been tried. The principal way in which technology assessments are made in this system, and probably in most governmental systems, is through the action of administrative agencies. There is no adversary process in administrative agencies.

Some of those present, however, saw serious problems with the adversary system. One speaker held:

> . . . anyone who has sat on a jury sometime must know that our judicial system is precisely designed to conceal the truth that any ordinary person would like to find out in order to reach a just verdict. To think that equitable and right and reasonable public policy can emerge from the deification of the judicial system is utter madness.

A professor of law felt that "one conspicuous weakness in the adversary process is that obviously and virtually by definition it doesn't always produce the truth." But he reaffirmed his faith in the process when he noted that "I don't know of any system that will produce the truth more frequently than the adversary process." A Government official who was not a lawyer felt that:

> . . . a defect of the adversary process is that it only chooses between alternatives presented to it. It doesn't make sufficient provision for proposing additional alternatives. So an adversary process may be a good process for what is called decision-making and a bad process for . . . policy formation.

THE NEED FOR CITIZEN PARTICIPATION

There was considerable discussion of the need for citizen participation and its value in the policy-making process. One participator felt the need arises from the fact that "there are unrepresented constituencies in the present process of deciding what direction various kinds of technology will go." Citizens currently feel alienated from the decision-making structure of both their local and national governments. Further,

> . . . much of the technology that the Federal Government has to do with is off in space or in the military. That is, it is out of the domain in which people live. That is why the Federal Government can make decisions on technology with a minimum interaction with the other political subdivisions of the country. Now industry is responsible for most of the technology that affects our daily lives. The problem is that the citizen does not have direct access through the political process to industry. He buys or doesn't buy, but on such a decentralized basis that he doesn't really affect the decision process.

The problem is an institutional one; at the moment there are no mechanisms which allow for a public voice (either directly or through inputs to governmental regulatory bodies) in society's choices about technology.

An industry representative pointed out that an aroused citizenry will always make itself heard in the political arena. To this, Mr. Mottur responded:

> . . . somehow citizens will express themselves. I have no doubt about that. What I am talking about here is the need to give them a mechanism where this can happen in a normal routine way.

A university researcher was concerned that citizens associations and conservation groups are "middle-class phenomena" so that any new citizens' action group would tend to "represent those who are already represented" and would still neglect the needs of those in the lower classes. But another participant disagreed:

> There are vast unrepresented groups in our society: middle-class and lower-class, black and white. I don't see anything in the proposal that suggests that it should be limited to white, middle-class groups.

The proposed caa's would, indeed, make it easier for anyone to become involved in influencing the policy process.

A Government official expressed concern about the proposal made in Mr. Mottur's paper and the value of widespread citizen participation in technology assessment:

> I think the position taken in the paper is fundamentally subversive because it reflects an attitude toward democracy which in turn reflects a pietistic mythology. I think the pietism lies in thinking that somehow or another the average man has something to say. I think the mythology lies in thinking that the government has the capability of initiating anything. When you combine pietism and mythology you are on the road to perdition.

He went on to describe the people who take an active concern with public issues as

> . . . very few in number. They are aberrant, they are extreme, they are unbalanced, and they are maniacal. They tend to be Johnny-one-notes, they tend to be the Carry Nations of the world. It doesn't make any difference what their issue is. Now the mechanism you propose would essentially take the nerve out of this Carry Nation mentality which is, in fact, the thing that has really made progress in our country. There is only a certain amount of emotional store in even . . . the concerned citizen. Your proposal is subversive in that it would dilute the

> mania which is in fact the main thing feeding democratic progress.

Mr. Mottur disagreed. His proposal, he felt, would certainly change the existing system of citizen participation. "Change," he felt, is a better description of the goal of the proposal than is "subversion." And he continued: "I don't believe that it would dilute the mania: I think you would enable the people with the mania to expand their effectiveness through being able to pull together larger groups." To the charge of pietism in his belief that the average man has something to say, Mr. Mottur responded that whether or not one believes that the average man has something to say is not the issue, "but I do make the assumption that in our society the average man has the right to have his say."

PARTICIPANTS

Thomas W. Adams, National Aeronautics and Space Administration

Bodo Bartocha, National Science Foundation

Raymond Bauer, Consultant to the White House for National Goals

Robert H. Becker, Kleinfeld and Kaplan Law Offices

Everett H. Bellows, Olin Corporation

Charles F. Bingman, National Aeronautics and Space Administration

James H. Brown, National Law Center, George Washington University

William F. Brown, Department of Transportation

Carl W. Bruch, Food and Drug Administration

John L. Buckley, Office of Science and Technology

William Carey, Arthur D. Little, Inc.

Richard A. Carpenter, Legislative Reference Service

Ira H. Cisin, George Washington University

Joseph F. Coates, National Science Foundation

Vary T. Coates, Program of Policy Studies, George Washington University

Joseph Cooper, Howard University

Harris Coulter, Consultant

Joseph D. Crumlish, National Bureau of Standards

Clarence H. Danhof, Program of Policy Studies, George Washington University

Charles L. Dunham, Staff, National Academy of Sciences

Steven Ebbin, Staff, National Academy of Sciences, National Academy of Engineering

Alfred J. Eggers, Jr., National Aeronautics and Space Administration

Lawrence A. Elgin, Consultant

Joseph V. Foa, George Washington University

James Gehrig, Staff, Aeronautical and Space Sciences Committee, U.S. Senate

Leo Goodman, United Auto Workers

Philip Gould, Institute for Defense Analyses

Irving Gray, Georgetown University

Harold P. Green, National Law Center, George Washington University

Leon Green, Jr., Lockheed Aircraft Corporation

Wayne R. Gruner, National Science Foundation

Walter A. Hahn, National Goals Research Staff

Charles Halpern, Center for Law and Social Policy

Irwin Halpern, National Goals Research Staff

John Hanessian, Jr., Program of Policy Studies, George Washington University

Robert B. Heller, George Washington University

Lawrence I. Hewes, III, J. Richard Taft Associates

Guenther Hintze, George Washington University

Isaac H. Hoover, Federal Aviation Agency

Robert E. Howard, Bureau of the Budget

Franklin P. Huddle, Legislative Reference Service

Alan H. Kaplan, Kleinfeld and Kaplan Law Offices

Norman Kaplan, Program of Policy Studies, George Washington University

Raphael G. Kasper, Program of Policy Studies, George Washington University

Frances O. Kelsey, Food and Drug Administration

Charles V. Kidd, Association of American Universities

David Kieffer, Chemical and Engineering News

William M. Kitzmiller, Staff of Representative Richard L. Ottinger

Vincent Kleinfeld, Kleinfeld and Kaplan Law Offices

Ezra Kohn, President's Commission on Consumer Interests

James L. Kunen, Eugene and Agnes Meyer Foundation

W. Henry Lambright, Syracuse University

Robert Lamson, National Science Foundation

Leonard Lederman, Batelle Memorial Institute

Joseph Luman, Staff, Military Operations Subcommittee, Committee on Government Operations, U.S. House of Representatives

James E. Mahoney, Program of Policy Studies, George Washington University

M. S. Malkin, NUS Corporation

Harold G. Mandel, George Washington University Medical School

Joseph B. Margolin, Program of Policy Studies, George Washington University

Rodney Markley, Ford Motor Company

William H. Marlow, George Washington University

Frank Marriott, Westinghouse Electric Corporation

Louis H. Mayo, Program of Policy Studies, George Washington University

Michael Michaelis, Arthur D. Little, Inc.

Arthur S. Miller, National Law Center, George Washington University

Clarence C. Mondale, George Washington University

Lawrence I. Moss, Staff, National Academy of Engineering

Ellis R. Mottur, Program of Policy Studies, George Washington University

Robert Multhauf, Smithsonian Institution

Arnold Nash, National Goals Research Staff

Harold Orlans, Brookings Institution

John B. Phelps, American Institute of Physics

Walter A. Radius, National Aeronautics and Space Administration

John Rasmussen, Yale University

Thomas Ratchford, Staff, Science and Astronautics Committee, U.S. House of Representatives

Leonard Rodberg, Institute for Policy Studies

S. McKee Rosen, George Washington University

Jean-Jacques Salomon, Organization for Economic Cooperation and Development

Leo Shubert, American University

Harvey B. Silverstein, Program of Policy Studies, George Washington University

Daniel Singer, Strasser, Spiegelberg, Fred, Frank and Kampelman

Nozer Singpurwalla, George Washington University

Francis B. Smith, National Aeronautics and Space Administration

Joel A. Snow, National Science Foundation

Henry Solomon, George Washington University

Richard Stephens, National Aeronautics and Space Administration

Robert L. Stern, Staff, National Academy of Engineering

Gabor Strasser, Office of Science and Technology

Charles S. Tidball, George Washington University Medical School

Duke C. Trexler, Staff, National Academy of Sciences

Lawrence H. Tribe, Staff, National Academy of Sciences

Joseph Tropea, George Washington University

E. Winslow Turner, Staff, Intergovernmental Relations Subcommittee, Committee on Government Operations, U.S. Senate

Ernest Weiss, National Transportation Safety Board

David H. West, Federal Aviation Agency

Charles W. Williams, National Goals Research Staff

Glen P. Wilson, Staff, Aeronautical and Space Sciences Committee, U.S. Senate

Clotaire Wood, National Aeronautics and Space Administration

Reuben E. Wood, George Washington University

James R. Wright, Airport Operators Council International

Philip Yeager, Staff, Science and Astronautics Committee, U.S. House of Representatives

ABOUT THE AUTHORS

RAPHAEL G. KASPER is a Research Scientist at The Program of Policy Studies in Science and Technology of The George Washington University.

ROBERT H. BECKER is an Associate, Kleinfeld and Kaplan Law Offices in Washington, D.C.

RICHARD A. CARPENTER was Senior Specialist in Science and Technology for the Science Policy Research Division of the Legislative Reference Service of the Library of Congress when his paper was prepared. He is currently Chief, Environmental Policy Division of the Congressional Research Service.

CLARENCE H. DANHOF was a Senior Staff Scientist at the Program of Policy Studies in Science and Technology of The George Washington University when his paper was prepared. He is currently Deputy Director, Science and Technology Project, Council of State Governments.

HAROLD P. GREEN is Professor of Law and Director of the Law, Science and Technology Program of The George Washington University National Law Center.

LEON GREEN, JR. was Director of Planning, Washington Area, Lockheed Aircraft Company when his paper was prepared. He is currently Executive Secretary of the Defense Science Board in the Office of the Director of Defense Research and Engineering in the Department of Defense.

FRANKLIN P. HUDDLE is Specialist in Science and Technology for the Science Policy Research Division of the Congressional Research Service of the Library of Congress.

ALAN H. KAPLAN is a Partner, Kleinfeld and Kaplan Law Offices in Washington, D.C.

CHARLES V. KIDD is Director of the Council on Federal Relations of the Association of American Universities. Until 1969 he was

Executive Secretary of the Federal Council for Science and Technology.

LOUIS H. MAYO is Vice President for Policy Studies and Special Projects and Director of The Program of Policy Studies in Science and Technology of The George Washington University.

ELLIS R. MOTTUR is Director of the Technology Assessment Implementation Project of The George Washington University.

ERNEST WEISS is Executive Director of the National Transportation Safety Board.